AQA

Physics for GCSE Combined Science: Trilogy

Third edition

Jim Breithaupt
Gary Calder
Editor: Lawrie Ryan

Message from AQA

This textbook has been approved by AQA for use with our qualification. This means that we have checked that it broadly covers the specification and we are satisfied with the overall quality. Full details of our approval process can be found on our website.

We approve textbooks because we know how important it is for teachers and students to have the right resources to support their teaching and learning. However, the publisher is ultimately responsible for the editorial control and quality of this book.

Please note that when teaching the *AQA GCSE Physics* or *AQA GCSE Combined Science: Trilogy* course, you must refer to AQA's specification as your definitive source of information. While this book has been written to match the specification, it cannot provide complete coverage of every aspect of the course.

A wide range of other useful resources can be found on the relevant subject pages of our website: www.aqa.org.uk.

OXFORD
UNIVERSITY PRESS

OXFORD
UNIVERSITY PRESS

Great Clarendon Street, Oxford, OX2 6DP, United Kingdom

Oxford University Press is a department of the University of Oxford.
It furthers the University's objective of excellence in research,
scholarship, and education by publishing worldwide. Oxford is a
registered trade mark of Oxford University Press in the UK and in
certain other countries

British Library Cataloguing in Publication Data
Data available

978019 835928 9

10 9 8 7 6

Paper used in the production of this book is a natural, recyclable
product made from wood grown in sustainable forests.
The manufacturing process conforms to the environmental regulations
of the country of origin.

Printed in Great Britain by Bell and Bain Ltd, Glasgow

MIX
Paper from
responsible sources
FSC
www.fsc.org   FSC® C007785

Acknowledgements

Jim Breithaupt wishes to acknowledge the support, advice, and
contributions he has received from Marie Breithaupt, Gary Calder,
and Darren Forbes, and from Emma Craig, Sadie Garratt, and their
colleagues at Oxford University Press.

Lawrie Ryan wishes to give thanks to the following people for their help
and support in producing this textbook. Each one added value to his
initial efforts: John Scottow, Annie Hamblin, Sadie Garratt, Emma-Leigh
Craig, Amie Hewish, and Sue Orwin.

AQA examination questions are reproduced by permission of AQA.

Index compiled by INDEXING SPECIALISTS (UK) Ltd., Indexing house,
306A Portland Road, Hove, East Sussex, BN3 5LP, United Kingdom.

COVER: Johnér / Offset
p2-3: Ezume Images/Shutterstock; **p4**: Leonid Andronov/Shutterstock; **p6**:
Sihasakprachum/Shutterstock; **p8(T)**: HconQ/Shutterstock; **p8(B)**: Jack
Guez/Getty Images; **p14(T)**: Sergey Novikov/Shutterstock; **p14(B)**: Syda
Productions/Shutterstock; **p18**: Martyn F. Chillmaid/Science Photo Library;
p19(T): Cordelia Molloy/Science Photo Library; **p19(B)**: Denis Kuvaev/
Shutterstock; **p20**: Twobee/Shutterstock; **p21**: Holbox/Shutterstock; **p25**:
Kzenon/Shutterstock; **p27**: Harvey Fitzhugh/Shutterstock; **p30**: Vladimir
Tronin/Shutterstock; **p32**: Ashley Cooper/Science Photo Library; **p34**: Lakov
Filimonov/Shutterstock; **p35(T)**: Zeljko Radojko/Shutterstock; **p35(B)**:
Chris Gallagher/Science Photo Library; **p36**: Paulrommer/Shutterstock;
p39: Julof90/iStockphoto; **p44-45**: Sommai/Shutterstock; **p51(T)**: Martyn
F. Chillmaid/Science Photo Library; **p51(B)**: Martyn F. Chillmaid/Science
Photo Library; **p61**: Ulkastudio/Shutterstock; **p62**: Raphael Gaillarde/Getty
Images; **p63**: Cordelia Molloy/Science Photo Library; **p66**: OUP Provided; **p70**:
Orchidflower/Shutterstock; **p72**: Wavebreakmedia/Shutterstock; **p96-97**:
Germanskydiver/Shutterstock; **p104(L)**: Everett Collection/Shutterstock;
p104(R): ARprofessionals.com.my/Shutterstock; **p106**: Nightman1965/
Shutterstock; **p108**: Susan Leggett/Shutterstock; **p112**: US Air Force/
Science Photo Library; **p114**: Brian Kinney/Shutterstock; **p128**: Paolo
Bona/Shutterstock; **p134-135**: Wang Song/Shutterstock; **p136**: HighTide/
Shutterstock; **p142**: Cuson/Shutterstock; **p150(T)**: DeepGreen/Shutterstock;
p150(B): Christian Delbert/Shutterstock; **p152**: US Air Force/Micaiah
Anthony/Science Photo Library; **p153**: Martyn F. Chillmaid/Science Photo
Library; **p154(T)**: AJ Photo/Hop Americain/Science Photo Library; **p154(B)**:
Wang Song/Shutterstock; **p155**: CNRI/Science Photo Library; **p158**: Awe
Inspiring Images/Shutterstock; **p174(T)**: Media Union/Shutterstock;
p174(B): Stephanie Bright; **p175(T)**: Galyna Andrushko/Shutterstock;
p175(C): Smileus/Shutterstock; **p175(B)**: Becris/Shutterstock; **p177**:
Omphoto/Shutterstock; **p178**: Mino Surkala/Shutterstock; **p181**:
FloridaStock/Shutterstock; **p183**: Royal Institution of Great Britain/Science
Photo Library; **p184**: Stefan Holm/Shutterstock; **p188**: Jerry Zitterman/
Shutterstock; **p189**: Peter Bernik/ Shutterstock; **p199**: Annto/Shutterstock;
p200: Jcjgphotography/Shutterstock; **p201**: David Iliff/Shutterstock; **p202**:
Hxdbzxy/Shutterstock; **p203**: Roman White/Shutterstock;

Header Photo_01: Ezume Images/Shutterstock;

Header Photo_02: Sommai/Shutterstock;

Header Photo_03: Germanskydiver/Shutterstock;

Header Photo_04: Wang Song/Shutterstock;

All artwork by Q2A Media

Contents

Required Practicals

Practical work is a vital part of physics, helping to support and apply your scientific knowledge, and develop your investigative and practical skills. In this Physics part of your AQA Combined Science: Trilogy course, there are eight required practicals that you must carry out. Questions in your exams could draw on any of the knowledge and skills you have developed in carrying out these practicals.

A Required practical feature box has been included in this student book for each of your required practicals. Further support is available on Kerboodle.

Required practicals	Topic
14 **Determining specific heat capacity.** Determine the specific heat capacity of a metal block of known mass by measuring the energy transferred to the block and its temperature rise, and using the equation for specific heat capacity.	P2.2
15 **Investigating resistance.** Set up circuits and investigate the resistance of a wire, and of resistors in series and parallel.	P4.2 P4.5
16 **Investigating electrical components.** Correctly assemble a circuit and investigate the potential difference–current characteristics of circuit components.	P4.3
17 **Calculating densities.** Measure the mass and volume of objects and liquids and calculate their densities using the density equation.	P6.1
18 **Investigate the relationship between force and extension for a spring.** Hang weights of known mass from a spring and, using the correct apparatus, measure the resulting extension. Use the results to plot a force-extension graph.	P10.5
19 **Investigate the relationship between force and acceleration.** Using a newton-metre, investigate the effect on the acceleration of an object of varying the force on it and of varying its mass.	P10.1
20 **Investigating plane waves in a ripple tank and waves in a solid.** Determine which apparatus are the most suitable for measuring the frequency, speed, and wavelength of waves in a ripple tank, and investigate waves on a stretched string.	P11.4
21 **Investigating infrared radiation.** Determine how the properties of a surface affect the amount of infrared radiation absorbed or radiated by the surface.	P12.2

Learning objectives

- Learning objectives at the start of each spread tell you the content that you will cover.
- Any objectives marked with the higher tier icon **H** are only relevant to those who are sitting the higher tier exams.

This book has been written by subject experts to match the new 2016 specifications. It is packed full of features to help you prepare for your course and achieve the very best you can.

Key words are highlighted in the text. You can look them up in the glossary at the back of the book if you are not sure what they mean.

The diagrams in this book are as important for your understanding as the text, so make sure you revise them carefully.

Synoptic link

Synoptic links show how the content of a topic links to other parts of the course. This will support you with the synoptic element of your assessment.

There are also links to the Maths skills for Physics chapter, so you can develop your maths skills whilst you study.

Practical

Practicals are a great way for you to see science in action for yourself. These boxes may be a simple introduction or reminder, or they may be the basis for a practical in the classroom. They will help your understanding of the course.

Required practical

These practicals have important skills that you will need to be confident with for part of your assessment. Your teacher will give you additional information about tackling these practicals.

Study tip

Hints giving you advice on things you need to know and remember, and what to watch out for.

Anything in the Higher Tier spreads and boxes must be learnt by those sitting the higher tier exam. If you will be sitting foundation tier, you will not be assessed on this content.

Higher

Go further!

Go further feature boxes encourage you to think about science you have learnt in a different context and introduce you to science beyond the specification. You do not need to learn any of the content in the Go further boxes.

Using maths
This feature highlights and explains the key maths skills you need. There are also clear step-by-step worked examples.

Summary questions

Each topic has summary questions. These questions give you the chance to test whether you have learnt and understood everything in the topic. The questions start off easier and get harder, so that you can stretch yourself.

The Literacy pen ✐ shows activities or questions that help you develop literacy skills.

Key points

Linking to the Learning objectives, the Key points boxes summarise what you should be able to do at the end of the topic. They can be used to help you with revision.

Any questions marked with the higher tier icon **H** are for students sitting the higher tier exams.

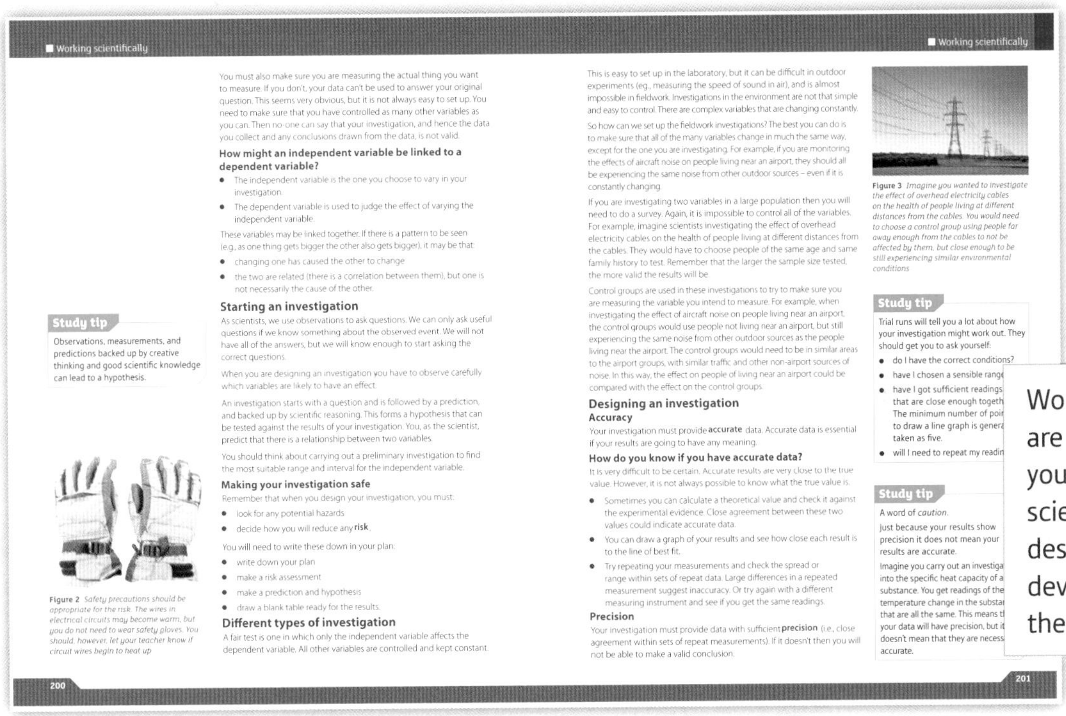

Figure 3 *Imagine you wanted to investigate the effect of overhead electricity cables on the health of people living at different distances from the cables. You would need to choose a control group using people far away enough from the cables to not be affected by them, but close enough to be still experiencing similar environmental conditions*

Working Scientifically skills are an important part of your course. The working scientifically section describes and supports the development of some of the key skills you will need.

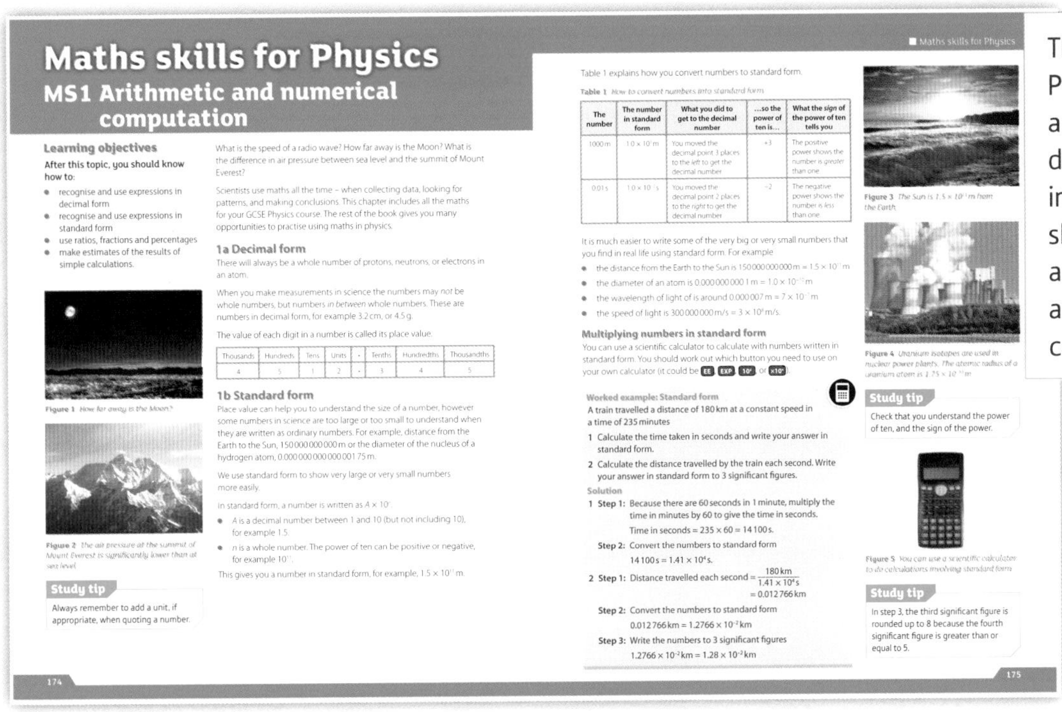

Maths skills for Physics
MS1 Arithmetic and numerical computation

Learning objectives
After this topic, you should know how to:
- recognise and use expressions in decimal form
- recognise and use expressions in standard form
- use ratios, fractions and percentages
- make estimates of the results of simple calculations.

Figure 1 *How far away is the Moon?*

Figure 2 *The air pressure at the summit of Mount Everest is significantly lower than at sea level*

The Maths skills for Physics chapter describes and supports the development of the important mathematical skills you will need for all aspects of your course. It also has questions so you can test your skills.

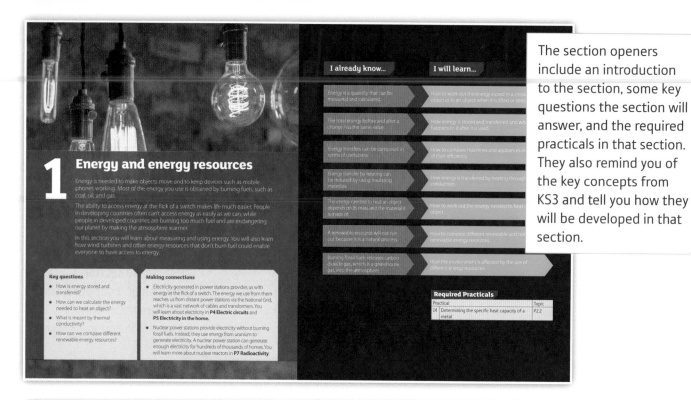

The section openers include an introduction to the section, some key questions the section will answer, and the required practicals in that section. They also remind you of the key concepts from KS3 and tell you how they will be developed in that section.

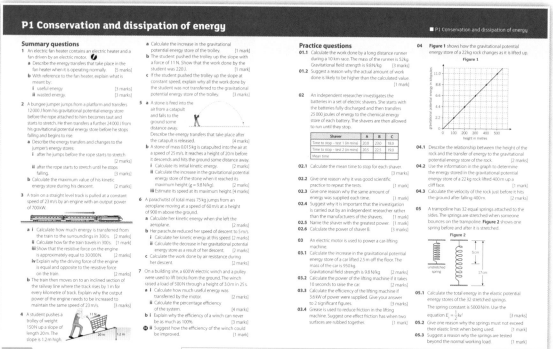

At the end of every chapter there are summary questions and practice questions. The questions test your literacy, maths, and working scientifically skills, as well as your knowledge of the concepts in that chapter. The practice questions can also call on your knowledge from any of the previous chapters to help support the synoptic element of your assessment.

There are also further practice questions at the end of the book to cover all of the content from your course.

Kerboodle

This book is also supported by Kerboodle, offering unrivalled digital support for building your practical, maths and literacy skills.

If your school subscribes to Kerboodle, you will find a wealth of additional resources to help you with your studies and revision:

- animations, videos, and revision podcasts
- webquests
- maths and literacy skills activities and worksheets
- on your marks activities to help you achieve your best
- practicals and follow-up activities
- interactive quizzes that give question-by-question feedback
- self-assessment checklists

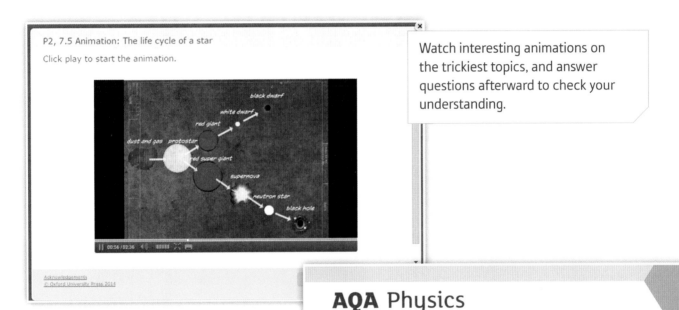

Watch interesting animations on the trickiest topics, and answer questions afterward to check your understanding.

If you are a teacher reading this, Kerboodle also has plenty of practical support, assessment resources, answers to the questions in the book, and a digital markbook along with full teacher support for practicals and the worksheets, which include suggestions on how to support and stretch your students. All of the resources that you need are pulled together into ready-to-use lesson presentations.

Check your own progress with the self-assessment checklists.

1 Energy and energy resources

Energy is needed to make objects move and to keep devices such as mobile phones working. Most of the energy you use is obtained by burning fuels, such as coal, oil, and gas.

The ability to access energy at the flick of a switch makes life much easier. People in developing countries often can't access energy as easily as we can, while people in developed countries are burning too much fuel and are endangering our planet by making the atmosphere warmer.

In this section you will learn about measuring and using energy. You will also learn how wind turbines and other energy resources that don't burn fuel could enable everyone to have access to energy.

Key questions

- How is energy stored and transferred?

- How can we calculate the energy needed to heat an object?

- What is meant by thermal conductivity?

- How can we compare different renewable energy resources?

Making connections

- Electricity generated in power stations provides us with energy at the flick of a switch. The energy we use from them reaches us from distant power stations via the National Grid, which is a vast network of cables and transformers. You will learn about electricity in **P4 Electric circuits** and **P5 Electricity in the home.**

- Nuclear power stations provide electricity without burning fossil fuels. Instead, they use energy from uranium to generate electricity. A nuclear power station can generate enough electricity for hundreds of thousands of homes. You will learn more about nuclear reactors in **P7 Radioactivity**.

I already know...

Energy is a quantity that can be measured and calculated.

The total energy before and after a change has the same value.

Energy transfers can be compared in terms of usefulness.

Energy transfer by heating can be reduced by using insulating materials.

The energy needed to heat an object depends on its mass and the material it is made of.

A renewable resource will not run out because it is a natural process.

Burning fossil fuels releases carbon dioxide gas, which is a greenhouse gas, into the atmosphere.

I will learn...

How to work out the energy stored in a moving object or in an object when it is lifted or stretched.

How energy is stored and transferred and what happens to it after it is used.

How to compare machines and appliances in terms of their efficiency.

How energy is transferred by heating through conduction.

How to work out the energy needed to heat an object.

How to compare different renewable and non-renewable energy resources.

How the environment is affected by the use of different energy resources.

Required Practicals

Practical		Topic
14	Determining the specific heat capacity of a metal	P2.2

Learning objectives

After this topic, you should know:

- the ways in which energy can be stored
- how energy can be transferred
- the changes in energy stores that happen when an object falls
- the energy transfers that happen when a falling object hits the ground without bouncing back.

On the move

Cars, buses, planes, and ships all use fuels as chemical energy stores. They carry their own fuel. Electric trains use energy transferred from fuel in power stations. Electricity transfers energy from the power station to the train.

Figure 1 *The French Train à Grande Vitesse electric train can reach speeds of more than 500 km/hour*

Energy can be stored in different ways and is transferred by heating, waves, an electric current, or when a force moves an object. Here are some examples:

- Chemical energy stores include fuels, foods, or the chemicals found in batteries. The energy is transferred during chemical reactions.
- Kinetic energy stores describe the energy an object has because it is moving.
- Gravitational potential energy stores are used to describe the energy stored in an object because of its position, such as an object above the ground.
- Elastic potential energy stores describe the energy stored in a springy object when you stretch or squash it.
- Thermal energy stores describe the energy a substance has because of its temperature.

Energy can be transferred from one store to another. In a torch, the torch's battery pushes a current through the bulb. This makes the torch bulb emit light, and also get hot (Figure 2).

When an electric kettle is used to boil water, the current in the kettle's heating element transfers energy to the thermal energy store of the water and the kettle.

When an object is thrown into the air, the object slows down as it goes up. Here, energy is transferred from the object's kinetic energy store to its gravitational potential energy store.

You can show the energy transfers by using a flow diagram:

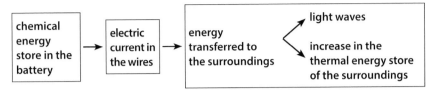

Figure 2 *Changes in energy stores in a torch lamp*

Energy transfers

When an object starts to fall freely, it speeds up as it falls. The force of gravity acting on the object causes energy to be transferred from its gravitational potential energy store to its kinetic energy store.

Look at Figure 3. It shows an object that hits the floor with a thud. All of the energy in its kinetic energy store is transferred by heating to the thermal energy store of the object and the floor, and by sound waves moving away from the point of impact. The amount of energy transferred by sound waves is much smaller than the amount of energy transferred by heating.

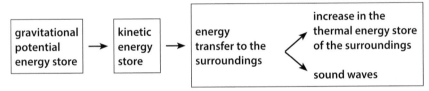

Figure 4 *An energy transfer diagram for an object when it falls and when it hits the ground*

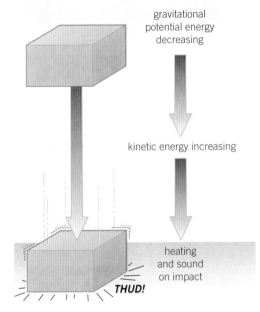

Figure 3 *An energetic drop. On impact, energy is transferred to the thermal energy store of the surroundings by heating and by sound waves*

1 Describe the changes to energy stores that take place when:
   a a ball falls in air   [2 marks]
   b an electric heater is switched on.   [2 marks]

2 **a** Name two different objects you could use to light a room if you have a power cut. For each object, describe the energy transfers and changes to energy stores that occur when it lights up the room. 🖊   [4 marks]
   b Which of the two objects in **a** is:
   i easier to obtain energy from?   [1 mark]
   ii easier to use?   [1 mark]

3 Describe the changes in energy stores of an electric train as it:
   a moves up a hill at constant speed   [2 marks]
   b approaches a station and brakes to a halt.   [2 marks]

4 Describe the changes in energy stores that take place when food is heated in a microwave oven. 🖊   [2 marks]

Key points

- Energy can be stored in a variety of different energy stores.
- Energy is transferred by heating, by waves, by an electric current, or by a force when it moves an object.
- When an object falls and gains speed, its store of gravitational potential energy decreases and its kinetic energy store increases.
- When a falling object hits the ground without bouncing back, its kinetic energy store decreases. Some or all of its energy is transferred to the surroundings – the thermal energy store of the surroundings increases, and energy is also transferred by sound waves.

P1.2 Conservation of energy

Learning objectives

After this topic, you should know:

- what conservation of energy is
- why conservation of energy is a very important idea
- what a closed system is
- how to describe the changes to energy stores in a closed system.

Figure 1 *Energy transfers on a roller coaster*

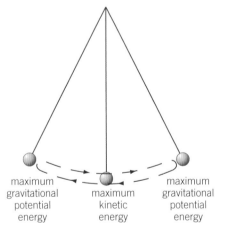

maximum gravitational potential energy   maximum kinetic energy   maximum gravitational potential energy

Figure 2 *A pendulum in motion. As the pendulum swings down and towards the centre, its gravitational potential energy store decreases as its kinetic energy store increases. As the pendulum moves upwards and away from the centre, its gravitational potential energy store increases as its kinetic energy store decreases*

At the funfair

Funfairs are very exciting places because changes to stores of energy happen quickly. As a roller coaster climbs an incline, its gravitational potential energy store increases. This energy is then transferred to other energy stores as the roller coaster races downwards.

As the roller coaster descends, its gravitational potential energy store decreases. Most of this energy is transferred to its kinetic energy store, which therefore increases. However, some energy is transferred to the thermal energy store of the surroundings by air resistance and friction, and some energy is transferred by sound waves.

Investigating pendulums

When changes to energy stores happen, does the total amount of energy stay the same? You can investigate this question with a simple pendulum.

Figure 2 shows a pendulum bob swinging from side to side.

As it moves towards the middle, energy is transferred by the force of gravity from its gravitational potential energy store to its kinetic energy store. So its gravitational potential energy store decreases and its kinetic energy store increases.

As it moves away from the middle, its kinetic energy store decreases and its gravitational potential energy store increases. If the air resistance on the bob is very small the bob will reach the same height on each side.

- Describe the changes to energy stores that take place in the bob when it goes from one side at maximum height to the other side at maximum height.

- Explain why it is difficult to mark the exact height the pendulum bob rises to. Suggest how you could make your judgement of height more accurate. 🖊

Conservation of energy

The pendulum in Figure 2 would probably keep on swinging for ever if it was in a vacuum because there would be no air resistance acting on it, and so no energy would be transferred from any of its energy stores. There would be no net change to the energy stored in the system. Because of this, it would be an example of a **closed system**.

A system is an object or a group of objects. Scientists have done lots of tests and have concluded that the total energy of a closed system is always the same before and after energy transfers to other energy stores within the closed system.

This important result is known as the principle of **conservation of energy**. It says that:

energy cannot be created or destroyed.

Energy can be stored in various ways. For example:

- when a rubber band is stretched, its elastic potential energy store increases
- when an object is lifted, its gravitational potential energy store is increased.

Bungee jumping

What energy transfers happen to a bungee jumper after jumping off the platform?

- When the rope is slack, energy is transferred from the gravitational potential energy store to the kinetic energy store as the jumper accelerates towards the ground due to the force of gravity.
- When the rope tightens, it slows the bungee jumper's fall. This is because the force of the rope reduces the speed of the jumper. The jumper's kinetic energy store decreases and the rope's elastic potential energy store increases as the rope stretches. Eventually the jumper comes to a stop – the energy that was originally in the kinetic energy store of the jumper has all been transferred into the elastic potential energy store of the rope.

After reaching the bottom, the rope recoils and pulls the jumper back up. As the jumper rises, the energy in the elastic potential energy store of the rope decreases and the bungee jumper's kinetic energy store increases until the rope becomes slack. After the rope becomes slack, and at the top of the ascent, the bungee jumper's kinetic energy store decreases to zero. The bungee jumper's gravitational potential energy store increases throughout the ascent.

The bungee jumper doesn't return to the original height. This is because some energy was transferred to the thermal energy store of the surroundings by heating as the rope stretched and then shortened again.

1 When a roller coaster gets to the bottom of a descent, describe the energy transfers and changes to energy stores that happen if:
   a the brakes are applied to stop it   [2 marks]
   b it goes up and over a second hill.   [3 marks]

2 **a** A ball dropped onto a trampoline returns to almost the same height after it bounces. Describe the energy transfers and changes to the energy stores of the ball from the point of release to the top of its bounce.   [3 marks]
   b Describe the energy stores of the ball at the point of release compared with its energy stores at the top of its bounce. [1 mark]
   c Describe how you would use the test in **a** to see which of three trampolines is the bounciest. 🖉   [5 marks]

3 One exciting fairground ride acts like a giant catapult. The capsule in which you are strapped is fired high into the sky by the rubber bands of the catapult. Explain the changes to the energy stores that take place in the ride as you move upwards.   [2 marks]

Bungee jumping
You can try out the ideas about energy transfers during a bungee jump using the experiment shown in Figure 3.

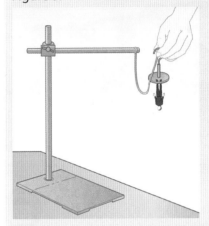

Figure 3 *Testing a bungee jump*

Safety: Make sure the stand is secure. Protect feet and bench from falling objects.

Key points

- Energy cannot be created or destroyed.
- Conservation of energy applies to all energy changes.
- A closed system is a system in which no energy transfers take place out of or into the energy stores of the system.
- Energy can be transferred between energy stores within a closed system. The total energy of the system is always the same, before and after, any such transfers.

P1.3 Energy and work

Learning objectives

After this topic, you should know:

- what work means in science
- how work and energy are related
- how to calculate the work done by a force
- what happens to work done to overcome friction.

Figure 1 *Working out*

Worked example

A builder pushed a wheelbarrow a distance of 5.0 m across flat ground with a force of 50 N. How much work was done by the builder?

Solution

work done = force applied × distance moved

$= 50 \, N \times 5.0 \, m$

$= \mathbf{250 \, J}$

Figure 2 *Pulling a lorry*

Working out

In a fitness centre or a gym, you have to work hard to keep fit. Lifting weights and pedalling on an exercise bike are just two ways to keep fit. Whichever way you choose to keep fit, you have to apply a force to move something. So the work you do causes a transfer of energy.

When an object is moved by a force, **work** is done on the object by the force. So the force transfers energy to the object. The amount of energy transferred to the object is equal to the work done on it. For example, to raise an object, you need to apply a force to it to overcome the force of gravity on it. If the work you do on the object is 20 J, the energy transferred to it must be 20 J. So its gravitational potential energy store increases by 20 J.

energy transferred = work done

The work done by a force depends on the size of the force and the distance moved. One joule of work is done when a force of one newton causes an object to move a distance of one metre in the direction of the force. To calculate the work done by a force when it causes displacement of an object, use this equation:

work done, W = force applied, F ×
(joules, J)   (newtons, N)
distance moved along the line of action of the force, s
(metres, m)

Superhuman force

Imagine pulling a lorry over 40 m. On level ground, a pull force of about 2000 N is needed. The work done by the pulling force is 80 kJ (= 2000 N × 40 m). Very few people can manage to pull with such force. Don't even try it, though. The people who can do it are very, very strong and have trained specially for it.

Doing work

Carry out a series of experiments to calculate the work done in performing the tasks below. Use a newton-meter to measure the force applied, and use a metre ruler to measure the distance moved.

1. Drag a small box a measured distance across a rough surface.

2. Repeat the test above with two rubber bands wrapped around the box (Figure 3).

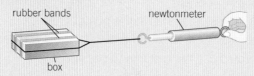

Figure 3 *At work*

- What is the resolution of your measuring instruments? Repeat your tests and comment on the precision of your repeat measurements. Can you be confident about the accuracy of your results?

Friction at work

Work done to overcome friction is mainly transferred to thermal energy stores by heating.

1 If you rub your hands together vigorously, they become warm. Your muscles do work to overcome the friction between your hands. The work you do is transferred as energy that warms your hands.

2 Brake pads on a vehicle become hot if the brakes are applied for too long. Friction between the brake pads and the wheel discs opposes the motion of the wheel. The force of friction does work on the brake pads and the wheel discs. As a result, energy is transferred from the kinetic energy store of the vehicle to the thermal energy store of the brake pads and the wheel discs. This makes them become hot and transfer energy by heating to the thermal energy store of the surrounding air.

3 Meteorites are small objects from space that enter the Earth's atmosphere and fall to the ground. As they pass through the atmosphere, friction caused by air resistance acts upon them. This results in energy being transferred from the meteorite's gravitational potential energy and kinetic energy stores to the meteorite's thermal energy store, causing the meteorite to heat up. If a meteorite becomes hot enough, it glows and becomes visible as a 'shooting star'. Very small objects can burn up completely. The surface of a space vehicle is designed to withstand the very high temperatures caused by this friction when it re-enters the Earth's atmosphere.

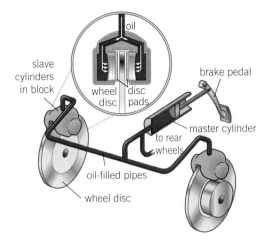

Figure 4 *Disc brakes*

1 a Describe what happens to the energy transferred:
   i by a rower rowing a boat [1 mark]
   ii by an electric motor used to raise a car park barrier. [1 mark]
  b Calculate how much work is done when a force of 3000 N pulls a truck through a distance of 50 m in the direction of the force. [1 mark]

2 A car is brought to a standstill when the driver applies the brakes.
  a Explain why the brake pads become warm. [2 marks]
  b The car travelled a distance of 20 metres after the brakes were applied. The braking force on the car during this time was 7000 N. Calculate the work done by the braking force. [1 mark]

3 a Calculate the work done when:
   i a force of 20 N makes an object move 4.8 m in the direction of the force [1 mark]
   ii an object of weight 80 N is raised through a height of 1.2 m. [1 mark]
  b When a cyclist brakes, his kinetic energy store is reduced from 1400 J to zero in a distance of 7.0 m. Calculate the braking force. [2 marks]

4 A student pushes a box at a steady speed a distance of 12 m across a level floor.
  a The student applied a horizontal force of 25 N to the box. Calculate the work done by the student. [2 marks]
  b Describe the energy transfers and changes to energy stores as the box moves. [3 marks]

Study tip

If you calculate the work done, this is equal to the energy transferred.

Key points

- Work is done on an object when a force makes the object move.
- Energy transferred = work done.
- Work done is $W = F\,s$ where F is the force and s is the distance moved (along the line of action of the force).
- Work done to overcome friction is transferred as energy to the thermal energy stores of the objects that rub together and to the surroundings.

P1.4 Gravitational potential energy stores

Learning objectives

After this topic, you should know:

- what happens to the gravitational potential energy stores of an object when it moves up or down
- why an object moving up has an increase in its gravitational potential energy store
- why it is easier to lift an object on the Moon than on the Earth
- how to calculate the change in gravitational potential energy of an object when it moves up or down.

Changes in gravitational potential energy stores

Every time you lift an object up, you do some work. Some of your muscles transfer energy from the chemical energy store in the muscle to the gravitational energy store of the object. In calculations we refer to the energy in this store as gravitational potential energy E_p.

The force you need to lift an object at constant velocity is equal and opposite to the gravitational force on the object. So the upward force you need to apply to it is equal to the object's weight. For example, you need a force of 80 N to lift a box of weight 80 N.

- When an object is moved upwards, the energy in its gravitational potential energy store increases. This increase is equal to the work done on it by the lifting force to overcome the gravitational force on the object.

- When an object moves down, the energy in its gravitational potential energy store decreases. This decrease is equal to the work done by the gravitational force acting on it.

The work done when an object moves up or down depends on:

1. how far it is moved vertically (its change of height)

2. its weight.

Using work done = force applied × distance moved in the direction of the force:

$$\begin{array}{ccc} \text{change in object's gravitational} & & \text{weight} & \text{change of height} \\ \text{potential energy store} & = & \text{(newtons, N)} \times & \text{(metres, m)} \\ \text{(joules, J)} & & & \end{array}$$

Worked example

A student of weight 300 N climbs on a platform that is 1.2 m higher than the floor. Calculate the increase in her gravitational potential energy store.

Solution

Increase of $E_p = 300\,\text{N} \times 1.2\,\text{m} = \mathbf{360\,J}$

Worked example

A 2.0 kg object is raised through a height of 0.4 m. Calculate the increase in the gravitational potential energy store of the object. The gravitational field strength of the Earth at its surface is 9.8 N/kg.

Solution

Gain of E_p = mass × gravitational field strength × height gain
= 2.0 kg × 9.8 N/kg × 0.4 m
= **7.8 J**

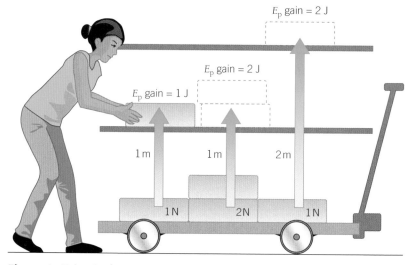

Figure 1 *Using joules*

Gravitational potential energy stores and mass

Astronauts on the Moon can lift objects much more easily than they can on the Earth. This is because the gravitational field strength on the Moon's surface is only about a sixth of the gravitational field strength on the Earth's surface.

You have previously learnt that the weight of an object in newtons is equal to its mass × the gravitational field strength. So, when an object is lifted or lowered, the change to its gravitational potential energy store is equal to its weight × its change of height. Therefore:

$$\begin{array}{c} \text{change of} \\ \text{gravitational} \\ \text{potential} \\ \text{energy store, } \Delta E_p \\ \text{(joules, J)} \end{array} = \begin{array}{c} \textbf{mass, } m \\ \text{(kilograms, kg)} \end{array} \times \begin{array}{c} \text{gravitational} \\ \textbf{field strength, } g \\ \text{(newtons per} \\ \text{kilogram, N/kg)} \end{array} \times \begin{array}{c} \textbf{change of} \\ \textbf{height, } \Delta h \\ \text{(metres, m)} \end{array}$$

1 a Describe the changes to the energy stores of a ball when it falls and rebounds without regaining its initial height. [4 marks]

b When a ball of weight 1.4 N is dropped from rest from a height of 2.5 m above a flat surface, it rebounds to a height of 1.7 m above the surface.

  i Calculate the total energy lost from the ball's energy stores when it reaches this maximum rebound height. [2 marks]

  ii Name two causes of the energy transfer. [2 marks]

2 A student of weight 450 N steps on a box of height 0.2 m.

a Calculate the increase in her gravitational potential energy store. [1 mark]

b Calculate the work done by the student if she steps on and off the box 50 times. [1 mark]

3 a A weightlifter raises a steel bar of mass 25 kg through a height of 1.8 m. Calculate the change to the gravitational potential energy store of the bar. The gravitational field strength at the surface of the Earth is 9.8 N/kg. [2 marks]

b The weightlifter then lowers the bar by 0.3 m and drops it so it falls to the ground. Assume that air resistance is unimportant. Calculate the change to its gravitational potential energy store in this fall. [2 marks]

4 You use energy when you hold an object stationary in your outstretched hand. Suggest what happens to the energy that must be supplied to your muscles to keep them contracted. 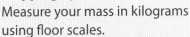 [3 marks]

P1.5 Kinetic energy and elastic energy stores

Learning objectives

After this topic, you should know:

- what the amount of energy in a kinetic energy store depends on
- how to calculate the amount of energy in a kinetic energy store
- what an elastic potential energy store is
- how to calculate the amount of energy in an elastic potential energy store.

The energy an object has because of its motion depends on its mass and speed. This energy is called kinetic energy.

Investigating kinetic energy stores

Figure 1 shows how you can investigate how the kinetic energy store of an object depends on its speed.

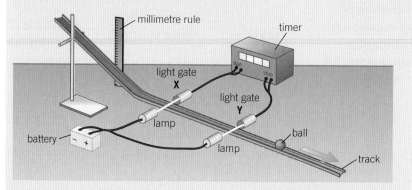

Figure 1 *Investigating changes in kinetic energy stores*

1   The ball is released on a slope from a measured height above the foot of the slope. You can calculate the decrease in its gravitational potential energy store by using the following equation:

change in gravitational potential energy store = mass × gravitational field strength × change in height.

Due to conservation of energy, this decrease in the gravitational potential energy store is matched by an equal increase in its kinetic energy store.

2   The ball is timed, using light gates, over a measured distance between X and Y after the slope.

- Explain why light gates improve the quality of the data you can collect in this investigation.

Investigating a catapult

Use rubber bands to catapult a trolley along a horizontal runway. Find out how the speed of the trolley depends on how much the catapult is pulled back before the trolley is released. For example, see if the distance needs to be doubled to double the speed. Figure 1 shows how the speed of the trolley can be measured.

Safety: Take care to ensure you do this safely. Protect your hands and feet, and the bench, from falling trolleys.

Table 1 shows some sample results.

Table 1 *Sample measurements for a ball of mass 0.5 kg*

Height drop to foot of slope in m	0.05	0.10	0.16	0.20
Initial kinetic energy of ball in J	0.25	0.50	0.80	1.00
Time to travel 1.0 m from X to Y in s	0.98	0.72	0.57	0.50
Average speed of ball between X and Y in m/s	1.02			2.00

Work out the speed of the ball between X and Y in each case. The first and last values have been worked out for you. Can you see a link between speed and height drop? The results show that the greater the height drop, the faster is the speed. So it can be said that the kinetic energy store of the ball increases if the speed increases.

The kinetic energy equation

Table 1 shows that when the height drop is increased by four times from 0.05 m to 0.20 m, the speed doubles. The height drop is directly proportional to the speed squared, or (speed)². Because the height drop is a measure of the ball's kinetic energy store, it can be said that the ball's kinetic energy store is directly proportional to the square of its speed.

The amount of energy in the kinetic energy store of an object can be calculated using the kinetic energy equation below:

kinetic energy, E_k = $\frac{1}{2}$ × mass, m × speed², v^2
(joules, J)   (kilograms, kg)   (metres per second, m/s)²

Using elastic potential energy

When you stretch a rubber band or a spring, the work you do is stored in it as elastic potential energy.

Figure 2 shows how the force F needed to stretch a spring varies with its extension e. The graph obeys the equation for **Hooke's Law** $F = k\,e$, where k is the **spring constant**.

For a spring stretched to an extension e, we can calculate the energy in its elastic potential energy store using the equation below:

elastic potential energy, $E_e = \frac{1}{2}$ × spring constant, k × extension², e^2

(joules, J)   (newtons per metre, N/m)   (metres, m)²

1 a Calculate the kinetic energy store of:
   i   a vehicle of mass 500 kg moving at a speed of 12 m/s [2 marks]
   ii a football of mass 0.44 kg moving at a speed of 20 m/s. [2 marks]
   b Calculate the velocity of a 500 kg vehicle with twice the kinetic energy store as calculated in **a i**. [3 marks]

2 a A catapult is used to fire an object into the air. Describe the energy transfers when the catapult is:
   i   stretched   [2 marks]   ii released.   [2 marks]
   b An object of weight 2.0 N fired vertically upwards from a catapult reaches a maximum height of 5.0 m. Calculate:
   i   the increase in the gravitational potential energy store of the object [2 marks]
   ii the speed of the object when it left the catapult. [4 marks]

3 A car moving at a constant speed has 360 000 J in its kinetic energy store. When the driver applies the brakes, the car stops in a distance of 100 m.
   a Calculate the force that stops the vehicle. [3 marks]
   b The speed of the car was 30 m/s when its kinetic energy store was 360 000 J. Calculate its mass. [3 marks]

4 A mobility aid to assist walking uses a steel spring to store energy when the walker's foot goes down, and it returns energy as the foot is lifted. The spring has a spring constant of 250 N/m. Calculate the elastic potential energy stored in the spring when its extension is 0.21 m. [2 marks]

Go further!

In Figure 2, the force F increases as the extension e is increased. The average force when the spring is extended to extension e is $\frac{1}{2}F$, where $F = ke$. Therefore, the energy stored in the spring = work done = average force × extension $= \frac{1}{2}F\,e = \frac{1}{2}ke^2$.

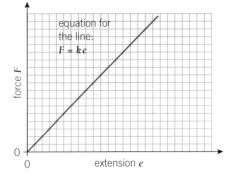

Figure 2 *Force versus extension for a spring. The spring constant k is the force per unit extension of the spring*

Key points

- The energy in the kinetic energy store of a moving object depends on its mass and its speed.
- The kinetic energy store of an object is $E_k = \frac{1}{2}mv^2$
- Elastic potential energy is the energy stored in an elastic object when work is done on the object.
- The elastic potential energy stored in a stretched spring is $E_e = \frac{1}{2}ke^2$, where e is the extension of the spring.

P1.6 Energy dissipation

Learning objectives

After this topic, you should know:

- what is meant by useful energy
- what is meant by wasted energy
- what eventually happens to wasted energy
- whether energy is still as useful after it is used.

Figure 1 *Using energy on a running machine*

a

drill bit

b

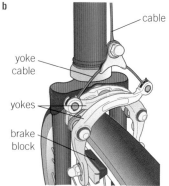

cable

yoke cable

yokes

brake block

Figure 2 *Friction in action: a Using a drill, b Braking on a bicycle*

Energy for a purpose

Where would people be without machines? Washing machines are used to clean your clothes. Machines in factories are used to make the things you buy. You might use exercise machines in a gym to keep yourself fit, and machines are used to get you from place to place.

A machine transfers energy for a purpose. Friction between the moving parts of a machine causes the parts to warm up. So, not all of the energy supplied to a machine is usefully transferred. Some of the energy is wasted.

- **Useful energy** is energy transferred to where it is wanted in the way that is wanted.

- **Wasted energy** is the energy that is not usefully transferred.

Whenever energy is transferred for a purpose in any system, some of the energy is transferred usefully. The rest is **dissipated** (spreads out) and may be stored in less useful ways. This energy is described as wasted energy because it is not transferred as useful energy. For example, when a jet plane takes off, its engines transfer energy from the chemical energy store in the fuel. Some of this energy increases the kinetic energy and the gravitational potential energy stores of the plane, which is the useful energy transfer pathway. The rest is wasted energy because some of it heats the plane and the surroundings and some is transferred to the surroundings by sound waves created by the engine vibrations.

Investigating friction

Friction in machines always causes energy to be wasted. Figure 2 shows two examples of friction in action.

In Figure 2a, friction acts between the drill bit and the wood. The bit becomes hot as it bores into the wood. Energy is transferred by an electric current to the thermal energy stores, heating up the drill bit and the wood.

When a bike rider brakes, friction acts between the brake blocks and the wheels (Figure 2b). Energy is wasted as it is transferred from the kinetic energy stores of the bike and the cyclist, to the thermal energy stores of the brake blocks and the wheels, which are heated by friction.

Friction at work

Next time you use a running machine, think about what happens to the energy transferred by your muscles. As you are exercising, energy is transferred to the thermal energy store of your muscles (so you get hot) and to the thermal energy store of the machine by the force you exert to overcome the friction in the machine.

Spreading out

- Wasted energy is dissipated (spreads out) to the surroundings, for example, the gears of a car get hot because of friction when the car is running. Here, energy is transferred from the kinetic energy store of the gear box to the thermal energy stores of the gear box and the surrounding air. The thermal energy stores of the gear box and the surrounding air therefore increase, as do their respective temperatures.

- Useful energy is eventually transferred to the surroundings too. For example, the useful energy supplied to turn the wheels of a car is eventually transferred from the kinetic energy stores of the wheels to the thermal energy stores of the tyres by heating – increasing the thermal energy stores of the tyres. This energy is then transferred to the thermal energy store of the road and surrounding air.

- Energy becomes less useful the more it spreads out. For example, the hot water from a central heating boiler in a building is pumped through pipes and radiators. The thermal energy store of the hot water decreases as it transfers energy by heating to the thermal energy stores of the radiators – heating the rooms in the building. But the energy supplied to heat these rooms will eventually be transferred to the surrounding air.

1 Copy and complete the table below.

Energy transfer by	Useful energy output	Wasted energy output
a An electric fan heater	warms the air and surrounding objects	
b A television		
c An electric kettle		
d Headphones		

[4 marks]

2 Describe what would happen, in terms of energy transfer and changes to energy stores, to:
   a a gear box that was insulated so it could not transfer energy by heating to the surroundings [3 marks]
   b the running shoes of a jogger if the shoes are well insulated [2 marks]
   c a blunt electric drill bit if you use it to drill into hard wood [2 marks]
   d the metal wheel discs of the brakes of a car when the brakes are applied. [2 marks]

3 **a** Describe the energy transfers and changes to energy stores of a pendulum as it swings from one side to the middle, and then to the opposite side. ✪ [4 marks]
   b Explain why a swinging pendulum eventually stops. [3 marks]

4 A freewheeling cyclist on a level road gradually stops moving. Describe the energy transfers and changes to the energy stores of the cyclist and the bicycle. ✪ [4 marks]

Key points

- Useful energy is energy in the place we want it and in the form we need it.
- Wasted energy is the energy that is not useful energy and is transferred by an undesired pathway.
- Wasted energy is eventually transferred to the surroundings, which become warmer.
- As energy dissipates (spreads out), it gets less and less useful.

P1.7 Energy and efficiency

Learning objectives

After this topic, you should know:

- what is meant by efficiency
- what is the maximum efficiency of any energy transfer
- how machines waste energy
- **❶** how energy transfers can be made more efficient.

When you lift an object, energy stored in your muscles is transferred to the gravitational potential energy store of the object. The amount transferred depends on the object's weight and how high you lift it.

- Weight is measured in newtons (N). The weight of a 1 kilogram object on the Earth's surface is about 10 N.
- Energy is measured in joules (J). The energy needed to lift a weight of 1 newton by a height of 1 metre is equal to 1 joule.

The energy supplied to the device is called the **input energy**. The useful energy transferred by the device is called the useful output energy.

Because energy cannot be created or destroyed:

Input energy (energy supplied, J) = useful output energy (useful energy transferred, J) + energy wasted (J)

For any device that transfers energy:

$$\text{efficiency} = \frac{\text{useful output energy transferred by the device (J)}}{\text{total input energy supplied to the device (J)}}$$

Efficiency

Efficiency can be written as a decimal number (that is always less than 1) or as a percentage.

For example, a light bulb with an efficiency of 0.15 would radiate 15 J of energy as light for every 100 J of energy you supply to it.

- Its efficiency (as a number) $= \frac{15}{100}$
  $= 0.15$
- Its percentage efficiency
  $= 0.15 \times 100 = \textbf{15\%}$

Worked example

An electric motor is used to raise an object. The object's gravitational potential energy store increases by 60 J when the motor is supplied with 200 J of energy by an electric current. Calculate the percentage efficiency of the motor.

Solution

Total energy supplied to the device = 200 J

Useful energy transferred by the device = 60 J

Percentage efficiency of the motor

$= \frac{\text{useful output energy transferred by the device}}{\text{total input energy supplied to the device}} \times 100$

$= \frac{60\,\text{J}}{200\,\text{J}} \times 100 = 0.30 \times 100\% = \textbf{30\%}$

Efficiency limits

No device can be more than 100% efficient, because you can never get more energy from a machine than you put into it.

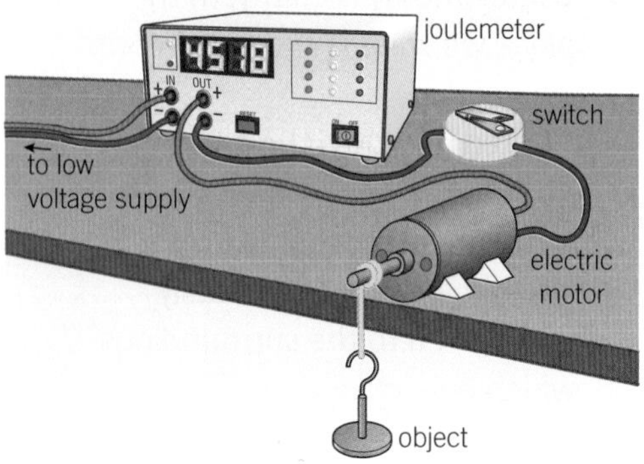

Investigating efficiency

Figure 1 shows how you can use an electric winch to raise a weight. You can use the joulemeter to measure the energy supplied.

- If you double the weight for the same increase in height, do you need to supply twice as much energy to do this task?

To calculate the change in the gravitational potential energy store of the weight use the following equation:

gravitational potential energy = weight in newtons × height increase in metres.

Figure 1 *An electric winch*

Use this equation and the joulemeter measurements to work out the percentage efficiency of the winch.

- Determine how the efficiency depends on the weight of the object being raised.

Plot a graph of your results, and use it to discuss how the efficiency of the winch changes with weight.

- Make some more measurements to find out whether lubricating the axle of the electric motor with a few drops of suitable oil makes the winch more efficient. Switch the motor off when you lubricate it.

Safety: Protect the floor and your feet. Stop the winch before the masses wrap round the pulley.

Improving efficiency

Table 1 *Increasing the efficiency of devices*

	Why devices waste energy	How to reduce the problem
1	Friction between the moving parts causes heating.	Lubricate the moving parts to reduce friction.
2	The resistance of a wire causes the wire to get hot when a current passes through it.	In circuits, use wires with as little electrical resistance as possible.
3	Air resistance causes a force on a moving object that opposes its motion. Energy transferred from the object to the surroundings by this force is wasted.	Streamline the shapes of moving objects to reduce air resistance.
4	Sound created by machinery causes energy transfer to the surroundings.	Cut out noise (e.g., tighten loose parts to reduce vibration).

1 a Compare the useful output energy by a machine and the input energy supplied to the machine. [1 mark]
   b Explain why the percentage efficiency of a machine can never be:
      i more than 100% [2 marks] ii equal to 100%. [4 marks]

2 An electric motor is used to raise a weight. When you supply 60 J of energy to the motor, the weight gains 24 J of gravitational potential energy. Calculate:
   a the energy wasted by the motor [1 mark]
   b the efficiency of the motor. [2 marks]

3 A machine is 25% efficient. If the total energy supplied to the machine is 3200 J, calculate how much useful energy can be transferred. [2 marks]

4 An electric fan heater contains a motor that blows hot air into the room. Describe the changes in the energy stores of the heater. ✪ [4 marks]

P1.8 Electrical appliances

Learning objectives

After this topic, you should know:

- how energy is supplied to your home
- why electrical appliances are so useful
- what most everyday electrical appliances are used for
- how to choose an electrical appliance for a particular job.

Synoptic links

You should know that electricity is not an energy store, but rather it is a flow of charge that transfers energy from one energy store to another. You will learn more about electricty in Chapter P4.

Everyday electrical appliances

The energy you use in your home is mostly supplied by electricity, gas, and oil. Although all three of these energy supplies can be used for cooking and heating, your electricity supply is vital because you use electrical appliances for so many purposes every day. The charge that flows through these appliances transfers energy to them, which they then transfer usefully. But some of the energy you supply to them is wasted.

Figure 1 *Electrical appliances – how many can you see in this photo?*

Table 1 *Comparing energy use in electrical appliances*

Appliance	Useful energy	Energy wasted
Light bulb	Light emitted from the glowing filament.	Energy transfer from the filament heating the surroundings.
Electric heater	Energy heating the surroundings.	Light emitted from the glowing element.
Electric toaster	Energy heating bread.	Energy heating the toaster case and the air around it.
Electric kettle	Energy heating water.	Energy heating the kettle itself.
Hairdryer	Kinetic energy of the air driven by the fan. Energy heating air flowing past the heater filament.	Sound of fan motor (energy heating the motor heats the air going past it, so is not wasted). Energy heating the hairdryer itself.
Electric motor	Kinetic energy of objects driven by the motor. Gravitational potential energy of objects lifted by the motor.	Energy heating the motor and energy transferred by the sound waves generated by the motor.

Clockwork radio

People without electricity supplies can now listen to radio programmes – thanks to the British inventor Trevor Baylis. In the early 1990s, he invented the clockwork radio. When you turn a handle on a clockwork radio, you wind up a clockwork spring in it, and increase the elastic potential energy of the spring. When the spring unwinds, energy from its elastic potential energy store is transferred to its kinetic energy store and it turns a small electric generator in the radio. It doesn't need batteries or mains electricity. So people in remote areas where there is no mains electricity can listen to their radios without having to walk miles for a replacement battery. But they do have to wind up the spring every time its store of energy has been used.

Figure 2 *Clockwork radios are now mass produced and sold all over the world*

Choosing an electrical appliance

You use electrical appliances for many purposes. Each appliance is designed for a specific purpose, and it should waste as little energy as possible. Suppose you were a rock musician at a concert. You would need appliances that transfer the variations in sound waves to electricity and then back to sound waves. But you wouldn't want the appliances to transfer lots of energy to the thermal energy store of the surroundings or themselves. See if you can spot some of these appliances in Figure 3.

Figure 3 *On stage*

1 Match each electrical appliance in the list below with the energy transfer A or B it is designed to bring about.

Energy transfer   **A**   Energy transferred by an electric current $\rightarrow$ energy transferred by sound waves

      B   Energy transferred by an electric current $\rightarrow$ kinetic energy store

   a Electric drill   [1 mark]
   b Food mixer   [1 mark]
   c Electric bell   [1 mark]

2 **a** Explain why a clockwork radio needs to be wound up before it can be used.   [2 marks]
   b Describe the changes in energy stores that take place in a clockwork radio when it is wound up and then switched on.   [2 marks]
   c Give one advantage and one disadvantage of a clockwork radio compared with a battery-operated radio.   [2 marks]

3 An electric dishwasher heats water and sprays the hot water at the dishes. It then pumps the water out.
   a Describe how energy is usefully used in the machine.   [1 mark]
   b Describe how energy is wasted by the machine.   [2 marks]

4 A laptop battery stores energy with an efficiency of 80% when it is recharged from a low-voltage power supply. When it is connected to the laptop, it transfers energy to the laptop with an efficiency of 60%.
   a Calculate the overall percentage efficiency of the laptop battery.   [2 marks]
   b Calculate the overall percentage of energy wasted.   [1 mark]

Key points

- Electricity and gas and/or oil supply most of the energy you use in your home.
- Electrical appliances can transfer energy in the form of useful energy at the flick of a switch.
- Uses of everyday electrical appliances include heating, lighting, making objects move (using an electric motor), and producing sound and visual images.
- An electrical appliance is designed for a particular purpose and should waste as little energy as possible.

P1.9 Energy and power

Learning objectives

After this topic, you should know:

- what is meant by power
- how to calculate the power of an appliance
- how to calculate the efficiency of an appliance in terms of power
- how to calculate the power wasted by an appliance.

Figure 1 *A lift motor*

Worked example

A motor transfers 10 000 J of energy in 25 s. Work out its power.

Solution

$$P = \frac{E}{t}$$

$$P = \frac{10\,000\,J}{25\,s} = 400\,W$$

When you use a lift to go up, a powerful electric motor pulls you and the lift upwards. The electric current through the lift motor transfers energy to the gravitational potential energy store of the lift when the lift goes up at a steady speed. You also get energy (from the electric current) transferred to the thermal energy store of the motor and the surroundings due to friction between the moving parts of the motor. In addition, energy is transferred to the thermal energy store of the surroundings by sound waves created by the lift machinery.

- The energy you supply to the motor per second is the **power** supplied to it.

- The more powerful the lift motor is, the faster it moves a particular load.

The more powerful an appliance is, the faster the rate at which it transfers energy.

The power of an appliance is measured in watts (W) or kilowatts (kW).

1 watt is equal to the rate of transferring 1 joule of energy in 1 second (i.e., 1 W = 1 J/s)

1 kilowatt is equal to 1000 watts (i.e., 1000 joules per second or 1 kJ/s).

You can calculate power using the equation:

$$\text{power, } P \text{ (watts, W)} = \frac{\text{energy transferred to appliance, } E \text{ (joules, J)}}{\text{time take for energy to be transferred, } t \text{ (seconds, s)}}$$

Power ratings

Table 1 shows some typical values of power ratings for different energy transfers.

Table 1

Appliance	Power rating
A torch	1 W
An electric light bulb	100 W
An electric cooker	10 000 W = 10 kW (where 1 kW = 1000 watts)
A railway engine	1 000 000 W = 1 megawatt (MW) = 1 million watts
A Saturn V space rocket	100 MW
A very large power station	10 000 MW
World demand for power	10 000 000 MW
A star like the Sun	100 000 000 000 000 000 000 000 MW

Orders of magnitude

The power ratings in Table 1 are called 'order of magnitude' values. This means that the values are estimates to the nearest power of ten.

The symbol ~ is used for an order of magnitude estimate. For example, if ten million people switch their electric kettles on at the same time (e.g., at half-time in a big TV football match), the jump in the demand for electric power would be ~ 10 000 MW (10^4 MW).

Muscle power

How powerful is a weightlifter?

A 30 kg dumbbell has a weight of 300 N. Raising it by 1 m would increase its gravitational potential energy store by 300 J. A weightlifter could lift it in about 0.5 seconds. The rate of energy transfer would be 600 J/s (= 300 J ÷ 0.5 s). So the weightlifter's power output would be about 600 W in total! The power output of two weightlifters could be compared by measuring how long they each take to lift the same weight through the same vertical height. The energy transferred is the same. So the one who takes the less time has a bigger power output.

Efficiency and power

For any appliance:

- its useful power out (or output power) is the useful energy per second transferred by it
- its total power in (or input power) is the energy per second supplied to it.

In Topic P1.7, you learnt that, for an appliance:

$$\text{efficiency} = \frac{\text{useful energy transferred by the device}}{\text{total energy supplied to it}} (\times 100)$$

Because power = energy per second transferred or supplied, this efficiency equation can be rewritten as

$$\textbf{efficiency} = \frac{\textbf{useful power out}}{\textbf{total power in}} (\times 100)$$

Wasted power

In any energy transfer, the energy wasted = the input energy supplied – the useful output energy. Because power is energy transferred per second:

power wasted = total power in – useful power out

1 **a** Determine which of the following is more powerful.
   i A torch bulb or a mains filament lamp. [1 mark]
   ii A 3 kW electric kettle or a 10 000 W electric cooker. [1 mark]

   b There are about 20 million occupied homes in England. If a 3 kW electric kettle was switched on in 1 in 10 homes at the same time, work out how much power would need to be supplied. [1 mark]

2 The total power supplied to a lift motor is 5000 W. In a test, the motor transfers 12 000 J of energy in 20 s to the gravitational potential energy store of the lift.
   a Calculate how much energy is supplied to the motor by the current through it in 20 s. [2 marks]
   b Calculate its efficiency in the test. [2 marks]

3 A machine has an input power rating of 100 kW. If the useful energy transferred by the machine in 50 seconds is 1500 kJ, calculate:
   a its output power in kilowatts [2 marks]
   b its percentage efficiency. [2 marks]

4 **a** Describe the energy transfers and changes to the energy stores of an electric hot-water shower when it is in operation. [5 marks]
   b A 12 kW electric shower is used 4 times in a day for 20 minutes each time. Calculate the energy supplied to it by electricity in one day. [2 marks]

Figure 2 *Muscle power*

Key points

- Power is rate of transfer of energy.
- The power of an appliance is $P = \frac{E}{t}$.
- efficiency of an appliance
$$= \frac{\text{useful power out}}{\text{total power in}} (\times 100)$$
- power wasted by an appliance = total power input − useful power output

P1 Conservation and dissipation of energy

Summary questions

1 An electric fan heater contains an electric heater and a fan driven by an electric motor.

  a Describe the energy transfers that take place in the fan heater when it is operating normally.   [5 marks]

  b With reference to the fan heater, explain what is meant by:

    i useful energy   [3 marks]

    ii wasted energy.   [3 marks]

2 A bungee jumper jumps from a platform and transfers 12 000 J from his gravitational potential energy store before the rope attached to him becomes taut and starts to stretch. He then transfers a further 24 000 J from his gravitational potential energy store before he stops falling and begins to rise.

  a Describe the energy transfers and changes to the jumper's energy stores:

    i after he jumps before the rope starts to stretch   [2 marks]

    ii after the rope starts to stretch until he stops falling.   [3 marks]

  b Calculate the maximum value of his kinetic energy store during his descent.   [2 marks]

3 A train on a straight level track is pulled at a constant speed of 23 m/s by an engine with an output power of 700 kW.

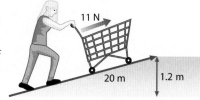

  a **i** Calculate how much energy is transferred from the train to the surroundings in 300 s.   [2 marks]

    ii Calculate how far the train travels in 300 s.   [1 mark]

    iii Show that the resistive force on the engine is approximately equal to 30 000 N.   [2 marks]

    iv Explain why the driving force of the engine is equal and opposite to the resistive force on the train.   [2 marks]

  b The train then moves on to an inclined section of the railway line where the track rises by 1 m for every kilometre of track. Explain why the output power of the engine needs to be increased to maintain the same speed of 23 m/s.   [3 marks]

4 A student pushes a trolley of weight 150 N up a slope of length 20 m. The slope is 1.2 m high.

  11 N

  20 m   1.2 m

  a Calculate the increase in the gravitational potential energy store of the trolley.   [1 mark]

  b The student pushed the trolley up the slope with a force of 11 N. Show that the work done by the student was 220 J.   [1 mark]

  c If the student pushed the trolley up the slope at constant speed, explain why all the work done by the student was not transferred to the gravitational potential energy store of the trolley.   [3 marks]

5 **a** A stone is fired into the air from a catapult and falls to the ground some distance away. Describe the energy transfers that take place after the catapult is released.   [4 marks]

  b A stone of mass 0.015 kg is catapulted into the air at a speed of 25 m/s. It reaches a height of 20 m before it descends and hits the ground some distance away.

    i Calculate its initial kinetic energy.   [2 marks]

    ii Calculate the increase in the gravitational potential energy store of the stone when it reached its maximum height ($g = 9.8$ N/kg).   [2 marks]

    iii Estimate its speed at its maximum height.   [4 marks]

6 A parachutist of total mass 75 kg jumps from an aeroplane moving at a speed of 60 m/s at a height of 900 m above the ground.

  a Calculate her kinetic energy when she left the aeroplane.   [2 marks]

  b Her parachute reduced her speed of descent to 5 m/s.

    i Calculate her kinetic energy at this speed.   [2 marks]

    ii Calculate the decrease in her gravitational potential energy store as a result of her descent.   [2 marks]

  c Calculate the work done by air resistance during her descent.   [2 marks]

7 On a building site, a 600 W electric winch and a pulley were used to lift bricks from the ground. The winch raised a load of 500 N through a height of 3.0 m in 25 s.

  a **i** Calculate how much useful energy was transferred by the motor.   [2 marks]

    ii Calculate the percentage efficiency of the system.   [4 marks]

  b **i** Explain why the efficiency of a winch can never be as much as 100%.   [3 marks]

    H **ii** Suggest how the efficiency of the winch could be improved.   [1 mark]

Practice questions

01.1 Calculate the work done by the engine of a radio controlled model car as it moves along a straight 1 km track. The driving force of the engine is 5.2 N. [3 marks]

01.2 Suggest a reason why the actual amount of work done is likely to be higher than the calculated value. [1 mark]

02 An independent researcher investigates the batteries in a set of electric shavers. She starts with the batteries fully discharged and then transfers 25 000 joules of energy to the chemical energy store of each battery. The shavers are then allowed to run until they stop.

Shaver	A	B	C
Time to stop – test 1 (in mins)	20.0	23.0	18.0
Time to stop – test 2 (in mins)	20.5	22.5	19.0
Mean time			

02.1 Calculate the mean time to stop for each shaver. [3 marks]

02.2 Give one reason why it was good scientific practice to repeat the tests. [1 mark]

02.3 Give one reason why the same amount of energy was supplied each time. [1 mark]

02.4 Suggest why it is important that the investigation is carried out by an independent researcher rather than the manufacturers of the shavers. [1 mark]

02.5 Name the shaver with the greatest power. [1 mark]

02.6 Calculate the power of shaver B. [3 marks]

03 An electric motor is used to power a car-lifting machine.

03.1 Calculate the increase in the gravitational potential energy store of a car lifted 2.5 m off the floor. The mass of the car is 950 kg. Gravitational field strength is 9.8 N/kg. [2 marks]

03.2 Calculate the power of the lifting machine if it takes 10 seconds to raise the car. [2 marks]

03.3 Calculate the efficiency of the lifting machine if 3.6 kW of power were supplied. Give your answer to 2 significant figures. [3 marks]

03.4 Grease is used to reduce friction in the lifting machine. Suggest one effect friction has when two surfaces are rubbed together. [1 mark]

04 **Figure 1** shows how the gravitational potential energy store of a 22 N rock changes as it is lifted up.

Figure 1

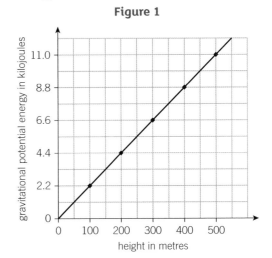

04.1 Describe the relationship between the height of the rock and the transfer of energy to the gravitational potential energy store of the rock. [2 marks]

04.2 Use the information in the graph to determine the energy stored in the gravitational potential energy store of a 22 N rock lifted 400 m up a cliff face. [1 mark]

04.3 Calculate the velocity of the rock just before it hits the ground after falling 400 m. [2 marks]

05 A trampoline has 32 equal springs attached to the sides. The springs are stretched when someone bounces on the trampoline. **Figure 2** shows one spring before and after it is stretched.

Figure 2

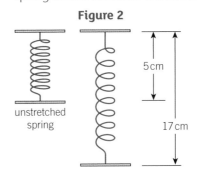

unstretched spring

5 cm

17 cm

05.1 Calculate the total energy in the elastic potential energy stores of the 32 stretched springs.

The spring constant is 5000 N/m. Use the equation $E_e = \frac{1}{2}ke^2$ [3 marks]

05.2 Give one reason why the springs must not exceed their elastic limit when being used. [1 mark]

05.3 Suggest a reason why the springs are tested beyond the normal working load. [1 mark]

Learning objectives

After this topic, you should know:

- which materials make the best conductors
- which materials make the best insulators
- how the thermal conductivity of a material affects the rate of energy transfer through it by conduction
- how the thickness of a layer of material affects the rate of energy transfer by conduction through it.

When you have a barbecue, you need to know which materials are good conductors and which ones are good insulators. If you can't remember, you're likely to burn your fingers!

Testing rods of different materials as conductors

The rods need to be the same width and length for a fair test. Each rod is coated with a thin layer of wax near one end. The uncoated ends are then heated together.

Look at Figure 1. The wax melts fastest on the rod that best conducts energy.

- Metals conduct energy better than non-metals.
- Copper is a better conductor than steel.
- Glass conducts better than wood.

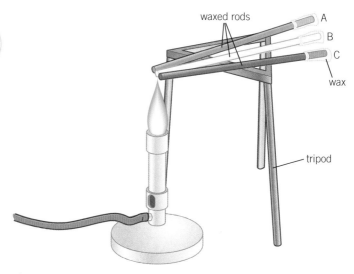

Figure 1 *Comparing conductors*

Testing sheets of materials as insulators

Use different materials to insulate identical cans (or beakers) of hot water.

When choosing your materials, consider which properties will make the materials good thermal insulators, for example, the thickness of the material or the colour of the material.

The volume of water and its temperature at the start should be the same.

Use a thermometer to measure the water temperature after a fixed time.

- Use your results to work out which is the best insulator.

The table below gives the results of comparing two different materials using the method above.

Material	Starting temperature in °C	Temperature after 300 s in °C
paper	40	32
felt	40	36

Safety: Take care if you are using very hot water.

Thermal conductivity

In Figure 1, each rod has the same temperature difference between its ends. Each rod is the same length and diameter. The energy transfer by conduction through a material depends on its **thermal conductivity**. The greater the thermal conductivity of a material, the more energy per second it transfers by conduction. So, in Figure 1, if

- A conducts better than C, and
- C conducts better than B, then
- the thermal conductivity of A is higher than the thermal conductivity of C, and
- the thermal conductivity of C is higher than the thermal conductivity of B.

Insulation matters

Materials that are good insulators are necessary to keep you warm in winter, whether you are at home or outdoors. Good insulators need to be materials that have low thermal conductivity, so energy transfer through them is as low as possible.

The energy transfer per second through a layer of insulating material depends on:

- the temperature difference across the material
- the thickness of the material
- the thermal conductivity of the material.

To reduce the energy transfer as much as possible in any given situation:

1 the thermal conductivity of the insulating material should be as low as possible

2 the thickness of the insulating layer should be as thick as is practically possible.

Figure 2 shows a layer of insulating material being fitted in the loft of a house. The insulating material chosen has a much lower thermal conductivity than the roof material. Several layers of this material fitted on the loft floor will reduce the energy transfer through the roof significantly. Insulating buildings is covered further in Topic P2.3.

Figure 2 *Insulating a loft. The air trapped between fibres makes fibreglass a good insulator*

1 **a** Explain why steel pans have handles made of plastic or wood.
[2 marks]

b Suggest which material, felt or paper, is the better insulator. Give a reason for your answer. [2 marks]

2 **a** Choose a material you would use to line a pair of winter boots. Give a reason for your choice of material. [2 marks]

b Describe how you could carry out a test on three different lining materials. Assume you have a thermometer, a stopwatch, and you can wrap the lining round a container of hot water. [3 marks]

3 Describe an investigation you would carry out to find out how the thickness of a layer of insulating material affects the energy transfer through it. [5 marks]

4 In Figure 1, A is a copper rod, B is a glass rod, and C is a steel rod. Determine which rod keeps its wax in the solid state for the longest time. Explain your answer. [2 marks]

Key points

- Metals are the best conductors of energy.
- Non-metal materials such as wool and fibreglass are the best insulators.
- The higher the thermal conductivity of a material, the higher the rate of energy transfer through it.
- The thicker a layer of insulating material, the lower the rate of energy transfer through it.

P2.2 Specific heat capacity

Learning objectives

After this topic, you should know:

- what is meant by the specific heat capacity of a substance
- how to calculate the energy changes that occur when an object changes temperature
- how the mass of a substance affects how quickly its temperature changes when it is heated
- how to measure the specific heat capacity of a substance.

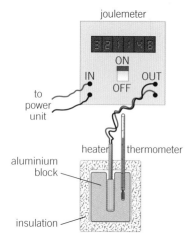

Figure 1 *Heating an aluminium block*

A car in strong sunlight can become very hot. A concrete block of equal mass would not become as hot. Metal heats up more easily than concrete. Investigations show that when a substance is heated, its temperature rise depends on:

- the amount of energy supplied to it
- the mass of the substance
- what the substance is.

The following results were obtained using two different amounts of water. They show that:

- heating 0.1 kg of water by 4 °C required an energy transfer of 1600 J
- heating 0.2 kg of water by 4 °C required an energy transfer of 3200 J

Using these results, you can say that:

- Increasing the temperature of 1.0 kg of water by 4 °C requires a transfer of 16 000 J of energy
- Increasing the temperature of 1.0 kg of water by 1 °C involves a transfer of 4000 J of energy.

More accurate measurements would give 4200 J per kg per °C for water. This is its **specific heat capacity**.

The specific heat capacity of a substance is the energy needed to raise the temperature of 1 kg of the substance by 1 °C.

The unit of specific heat capacity is the joule per kilogram degree Celsius (J/kg °C).

For a known change of temperature of a known mass of a substance:

| **energy transferred, ΔE** (joules, J) | = | **mass, m** (kilograms, kg) | × | **specific heat capacity, c** (joule per kilogram per degree Celsius, J/kg °C) | × | **temperature change, $\Delta \theta$** (degree Celsius, °C) |

The energy transferred to the substance increases the thermal energy store of the substance by an equal amount.

To find the specific heat capacity of a substance, rearrange the above equation into the form:

$$c = \frac{\Delta E}{m \, \Delta \theta}$$

Measuring specific heat capacity

Use the arrangement shown in Figure 1 to heat a metal block of known mass. You will need to use a thermometer and a top-pan balance.

Use the energy meter (or joulemeter) to measure the energy supplied to the block. Use the thermometer to measure its temperature rise.

What changes to energy stores occur as result of the transfer of energy to the block?

To find the specific heat capacity of aluminium, insert your measurements into the equation:

$$c = \frac{\Delta E}{m\,\Delta\theta}$$

Replace the block with an equal mass of water in a suitable container. Measure the temperature rise of the water when the same amount of energy is supplied to it by the heater.

Your results should show that aluminium heats up more quickly than water.

Safety: Wear eye protection and take care with a hot immersion heater.

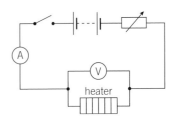

Figure 2 Circuit diagram

Storage heaters

A storage heater uses electricity at night (off-peak) to heat special bricks or concrete blocks in the heater. Energy transfer from the bricks keeps the room warm. The bricks have a high specific heat capacity, so they store lots of energy. They warm up slowly when the heater element is on, and cool down slowly when it is off.

Table 1 The specific heat capacity for some other substances

Substance	water	oil	aluminium	iron	copper	lead	concrete
Specific heat capacity in J/kg °C	4200	2100	900	390	385	130	850

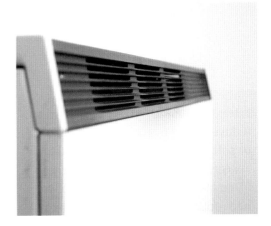

Figure 3 A storage heater

1 A small bucket of water and a large bucket of water are left in strong sunlight. Which one warms up faster? Give a reason for your answer. [2 marks]

2 Use the information in Table 1 above to answer this question.
  a Explain why a mass of lead heats up more quickly than an equal mass of aluminium. [2 marks]
  b Calculate the energy needed:
    i to raise the temperature of 0.20 kg of aluminium from 15 °C to 40 °C [2 marks]
    ii to raise the temperature of 0.40 kg of water from 15 °C to 40 °C [2 marks]
    iii to raise the temperature of 0.40 kg of water in an aluminium container of mass 0.20 kg from 15 °C to 40 °C. [3 marks]
  c A copper water tank of mass 20 kg contains 150 kg of water at 15 °C. Calculate the energy needed to heat the water and the tank to 55 °C. [5 marks]

3 Name *two* ways in which a storage heater differs from a radiant heater. [2 marks]

4 Design an experiment to measure the specific heat capacity of oil using the arrangement in Figure 1.  [6 marks]

Key points

- The specific heat capacity of a substance is the amount of energy needed to change the temperature of 1 kg of the substance by 1 °C.
- Use the equation $\Delta E = m\,c\,\Delta\theta$ to calculate the energy needed to change the temperature of mass m by $\Delta\theta$.
- The greater the mass of an object, the more slowly its temperature increases when it is heated.
- To find the specific heat capacity c of a substance, use a joulemeter and a thermometer to measure ΔE and $\Delta\theta$ for a measured mass m, then use $c = \frac{\Delta E}{m\,\Delta\theta}$.

P2.3 Heating and insulating buildings

Learning objectives

After this topic, you should know:

- how homes are heated
- how you can reduce the rate of energy transfer from your home
- what cavity wall insulation is.

Reducing the rate of energy transfers at home

Houses are heated by electric or gas heaters, oil or gas central heating systems, or by solid fuel in stoves or in fireplaces. Whichever form of heating you have in your home, the heating bills can be expensive. When your home heating system is transferring energy into your home to keep you warm, energy is also transferring to the surroundings outside your home. Figure 1 shows some of the measures that can be taken to reduce the rate of energy transfer from a home, and so reduce home heating bills.

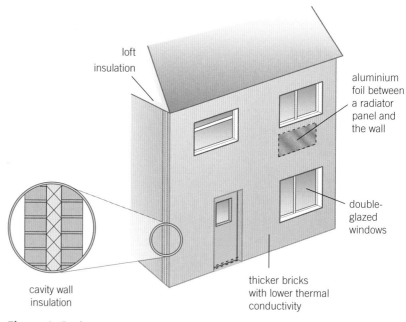

loft insulation

aluminium foil between a radiator panel and the wall

double-glazed windows

cavity wall insulation

thicker bricks with lower thermal conductivity

Figure 1 *Saving money*

Go further!

A duvet is a bed cover filled with 'down' or soft feathers, or some other suitable insulator such as wool. Because the filling material traps air, a duvet on a bed reduces the rate at which energy is transferred from you as you sleep. The tog rating of a duvet depends on the thickness of the material and on its thermal conductivity. It tells you how effective it is as an insulator. The thicker the material, or the lower its thermal conductivity, the better it is as an insulator, and so the higher its tog rating.

- Loft insulation such as fibreglass reduces the rate of energy transfer through the roof. Fibreglass is a good insulator. The air between the fibres also helps to reduce the rate of energy transfer by conduction. The greater the number of layers of insulation, the thicker the insulation will be. So the rate of energy transfer through the roof will be less.

- Cavity wall insulation reduces the rate of energy transfer through the outer walls of the house. The cavity of an outer wall is the space between the two layers of brick that make up the wall. The insulation is pumped into the cavity. It is a better insulator than the air it replaces. It traps the air in small pockets, reducing the rate of energy transfer by conduction.

- Aluminium foil between a radiator panel and the wall reflects radiation away from the wall and so reduces the rate of energy transfer by radiation.

- Double-glazed windows have two glass panes with dry air or a vacuum between the panes. The thicker the glass and the lower its thermal conductivity is, the slower the rate of transfer of energy through it by conduction will be. Dry air is a good insulator, so it reduces the rate of energy transfer by conduction. A vacuum also prevents energy transfer by convection.

- If the external walls of a warm building have thicker bricks and lower thermal conductivity, the rate of transfer of energy from the inside of the building to the outside will be lower and the cost of heating will be less.

Solar panels

Heating a home using electricity or gas can be expensive. Solar panels absorb infrared radiation from the Sun and are used to generate electricity directly (solar cell panels) or to heat water directly (solar heating panels). In the northern hemisphere, a solar panel is usually fitted on a roof that faces south so that it absorbs as much infrared radiation from the Sun as possible.

Synoptic link

You will learn more about solar panels in Topic P3.3.

1 **a** Explain why cavity wall insulation is better than air in the cavity between the walls of a house. [2 marks]
   b Explain why fixing aluminium foil to the wall behind a radiator reduces energy transfer through the wall. [2 marks]

2 Some double-glazed windows have a plastic frame and a vacuum between the panes.
   a Explain why a plastic frame is better than a metal frame. [2 marks]
   b Explain why a vacuum between the panes is better than air. [1 mark]

3 Two manufacturers advertise double-glazed windows of the same size and with dry air between the panes at the same price, but with a different gap width between the glass panes. Explain which one you would choose. ✍ [2 marks]

4 A manufacturer of loft insulation claimed that each roll of loft insulation would save £10 per year on fuel bills. A householder bought six rolls of the loft insulation at £15 per roll and paid £90 to have the insulation fitted in her loft.
   a Calculate how much it cost to buy and install the loft insulation. [2 marks]
   b Calculate what the saving each year would be on fuel bills. [1 mark]

Key points

- Electric and/or gas heaters and gas or oil-fired central heating or solid-fuel stoves are used to heat houses.
- The rate of energy transfer from houses can be reduced by using:
  - loft insulation
  - cavity wall insulation
  - double-glazed windows
  - aluminium foil behind radiators
  - external walls with thicker bricks and lower thermal conductivity.
- Cavity wall insulation is insulation material that is used to fill the cavity between the two brick layers of an external house wall.

P2 Energy transfer by heating

Summary questions

1 Explain why:
  a An electric iron has a plastic handle and a metal base. [4 marks]
  b Fitting layers of insulation in the loft of a house saves money. [2 marks]
  c Wrapping a block of ice cream in paper helps to stop it melting. [2 marks]

2 A heat sink is a metal plate or clip fixed to an electronic component to stop it overheating.

  a When the component becomes hot, how does energy transfer from:
    i where the component is in contact with the plate to the rest of the plate? [1 mark]
    ii the plate to the surroundings? [1 mark]
  b Describe the purpose of the metal fins on the plate. [2 marks]
  c Heat sinks are made from metals such as copper or aluminium. Copper is approximately three times more dense than aluminium, and its specific heat capacity is about twice as large. Explain how these physical properties are relevant to the choice of whether or not to use copper or aluminium for a heat sink in a computer. [4 marks]

3 **a** Explain why woolly clothing is very effective at keeping people warm in winter. [3 marks]
  b Wearing a hat in winter is a very effective way of keeping your head warm. Describe how a hat helps to reduce the rate of energy transfer from your head. [3 marks]

4 In an experiment to test the effectiveness of insulation material, hot water was poured into a test tube and the water temperature was measured at intervals while it cooled. The test was then repeated with the same volume of water, with the tube wrapped in the insulating material. The measurements are shown in Table 1.

Table 1

Time (seconds)	Temperature (°C)	
	Unwrapped tube	Wrapped tube
0	70	70
50	64	67
100	60	64.5
150	57	62
200	55	60
250	53.5	58.5

  a Plot graphs of temperature against time for both the wrapped and the unwrapped tubes. Label your graphs 'unwrapped' and 'wrapped'. [4 marks]
  b Estimate how long each tube took to cool from 70°C to 65°C. [2 marks]
  c Explain how your estimate in **b** for the wrapped tube would have differed if the test had been carried out with the test tube wrapped:
    i in two layers of the same insulation [3 marks]
    ii in a layer of insulation of the same thickness as the original material but with twice the thermal conductivity. [3 marks]
  d Explain why it was important in each test:
    i to use the same volume of water [2 marks]
    ii to start the timing at the same temperature. [2 marks]

5 A meteorite loses about 60 MJ/kg of energy from its kinetic and gravitational potential energy stores when it falls to the ground from space. The specific heat capacity of a meteorite is about 400 J/kg°C.
  a Estimate its maximum temperature rise if just 1% of the kinetic and gravitational potential energy is transferred as thermal energy of the meteorite. [4 marks]
  b The melting point is about 2500°C. Evaluate whether or not the 1% assumption in **a** is realistic. [2 marks]

6 A 5 kW electric shower heats the water flowing through it from 15°C to 40°C when the water flow rate is 1.5 kg per minute.
  a Calculate the energy per second used to heat the water. The specific heat capacity of water is 4200 J/kg°C. [2 marks]
  b Calculate the percentage efficiency of the shower heater. [3 marks]

7 Water in a plastic jug was put in a fridge where it cooled from 18°C to 4°C in 520 seconds. The mass of water in the jug was 0.85 kg.
The specific heat capacity of water is 4200 J/kg°C.
  a Calculate the energy transferred from the thermal energy store of the water. [2 marks]
  b Estimate an average value for the rate of transfer of energy from the water. [2 marks]

Practice questions

01 A student had read about an outdoor ice container. The article described ways of slowing down the rate at which the ice melts on hot summer days. She decided to investigate using the apparatus in **Figure 1**.

Figure 1

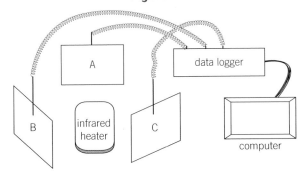

A is copper
B is glass
C is plastic

01.1 Why did the student use an infrared heater in the investigation? [1 mark]

01.2 Name two control variables in the investigation. [2 marks]

01.3 The student used a temperature probe attached to a data logger instead of a thermometer. Suggest how this improved the investigation. [2 marks]

01.4 A politician has suggested glaciers should be covered in insulation to slow down their rate of melting. Do you agree with this suggestion? Explain your answer [2 marks]

02 **Table 1** shows the thermal conductivity of three metals used in the manufacture of saucepans.

Table 1

Metal	Thermal conductivity in W/m²°C
copper	380
stainless Steel	54
aluminium	250

02.1 Choose one metal for the base of a saucepan that would give the best thermal efficiency. Give a reason for your choice. [2 marks]

02.2 Describe how to check a saucepan is hot without touching it or using a thermometer. [2 marks]

03.1 A hot water bottle made of rubber is filled with 0.65 kg of hot water. The temperature of the water is 90°C. Calculate the temperature of the hot water bottle after 163 800 J of energy are transferred during the night. Specific heat capacity of water is 4200 J/kg°C [3 marks]

03.2 A new type of bed warmer is sealed, filled with polymer gel, and heated using an electric insert. The bed warmer can control the temperature of the polymer gel. Suggest two advantages of using the new bed warmer rather than a traditional hot water bottle. [2 marks]

04 A student investigated the insulation properties of two materials, **A** and **B**. The apparatus she used is shown in Figure 2.

Figure 2

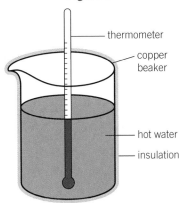

Her method was as follows:

1 Wrap a 2 cm layer of material **A** around the beaker.
2 Fill the beaker with 200 ml of hot water and record the temperature of the water.
3 Record the temperature of the water after 10 minutes.
4 Wrap a second 2 cm layer of material **A** around the beaker.
5 Repeat stages 2 and 3.
6 Replace material **A** with material **B** and repeat stages 1–5.

Table 2 shows the results of the investigation.

Table 2

Material	Number of layers	Water temperature at the start, in °C	Water temperature after 10 mins, in °C
A	1	82.5	66.0
B	1	83.0	71.5
A	2	81.5	72.0
B	2	75.0	67.5

04.1 Calculate the temperature change for each test. [2 marks]

04.2 Which material was the better insulator? Give a reason for your answer. [2 marks]

04.3 Suggest **two** ways the student could have improved the investigation. [2 marks]

Learning objectives

After this topic, you should know:

- how most of your energy demands are met today
- what other energy resources are used
- how nuclear fuels are used in power stations
- what other fuels are used to generate electricity.

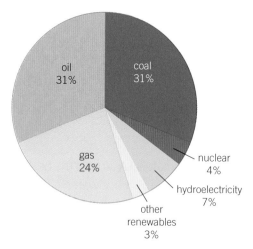

Figure 1 *World energy demand and sources of energy in 2013*

Figure 2 *Using biofuel to generate electricity*

Most of the energy you use comes from burning fossil fuels, mostly gas or oil or coal. The energy in homes, offices, and factories is mostly supplied by gas or by electricity generated in coal or gas-fired power stations. Oil is needed to keep road vehicles, ships, and aeroplanes moving. Burning one kilogram of fossil fuel releases about 30 million joules of energy. You use about 5000 joules of energy each second, which is about 150 thousand million joules each year. But because of the inefficiencies in how energy is distributed and used, a staggering 10 000 kg of fuel is used each year to supply the energy needed just for you!

Figure 1 shows how the global demand for energy is met. Fossil fuels are extracted from underground or under the sea bed and then transported to oil refineries and power stations. Much of the electricity you use is generated in fossil-fuel power stations. Instead of fossil fuels, some power stations use biofuels or nuclear fuel. Fossil fuels and nuclear fuel are non-renewable because they can not be replaced. As you will learn in Topic P3.4, their use is causing major environmental problems and increasing the levels of greenhouse gases, such as carbon dioxide, in the atmosphere. Some of the electricity you use is from renewable energy resources such as wind energy, hydroelectricity and solar energy, which you'll learn more about in Topics P3.2 and P3.3.

Inside a power station

In coal- or oil-fired power stations, and in most gas-fired power stations, the burning fuel heats water in a boiler. This produces steam. The steam drives a turbine that turns an electricity generator. Coal, oil, and gas are fossil fuels, which are fuels that come from long-dead animals and plants.

Biofuels

Methane gas can be collected from cows or animal manure, from sewage works, decaying rubbish, and other sources. It can be used in small gas-fired power stations. Methane from these sources is an example of a biofuel.

A **biofuel** is any fuel taken from living or recently living organisms. Animal waste is an example of a biofuel. Biofuels can be used instead of fossil fuel in modified engines for transport and in generators at power stations. Biodiesel uses waste vegetable oil and plants such as rapeseed. Other examples of biofuels are ethanol (from fermented sugar cane), straw, nutshells, and woodchip.

A biofuel is:

- **renewable** because its biological source either regrows (vegetation) or is continually produced (sewage and rubbish). This means it is used at the same rate that it is replaced.
- **carbon-neutral** because, in theory, the carbon that the living organism takes in from the atmosphere as carbon dioxide can balance the amount that is released when the biofuel is burnt.

Nuclear power

Nuclear fuel takes energy from atoms. Figure 3 shows that every atom contains a positively charged **nucleus** surrounded by electrons.

The fuel in a nuclear power station is uranium (or plutonium). The uranium fuel is in sealed cans in the core of the reactor. The nucleus of a uranium atom is unstable and can split in two. Energy is transferred from the nucleus when this happens. Because there are lots of uranium atoms in the core, it becomes very hot.

The energy of the core is transferred by a fluid (called the coolant) that is pumped through the core.

- The coolant is very hot when it leaves the core. It flows through a pipe to a heat exchanger, then back to the **reactor core**.

- The energy transferred by the coolant is used to turn water into steam in the heat exchanger. The steam drives turbines that turn electricity generators.

Table 1 *Comparing nuclear power and fossil fuel power*

	Nuclear power station	**Fossil fuel power station**
Fuel	Uranium or plutonium	Coal, oil, or gas
Energy released per kg of fuel	≈ 300 000 MJ (= about 10 000 × energy released per kg of fossil fuel)	≈ 30 MJ
Waste	Radioactive waste that needs to be stored for many years	Non-radioactive waste
Greenhouse gases, e.g., carbon dioxide	No – because uranium releases energy without burning	Yes – because fossil fuels produce gases such as carbon dioxide when they burn

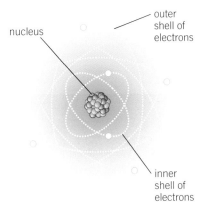

Figure 3 *The structure of the atom*

1 **a i** Name the types of power stations that release carbon dioxide into the atmosphere. [1 mark]
   ii Name the type of power station that does not release carbon dioxide into the atmosphere. [1 mark]
   b Nuclear fuel releases about 10 000 times as much energy as the same mass of fossil fuel. Give one disadvantage of nuclear fuel compared with other types of fuel. [2 marks]

2 **a** Give one advantage and one disadvantage of:
   i an oil-fired power station compared with a nuclear power station [1 mark]
   ii a gas-fired power station compared with a coal-fired power station. [1 mark]
   b Look at Table 1.
   Calculate how many kilograms of fossil fuel would give the same amount of energy as 1 kilogram of uranium fuel. [1 mark]

3 **a** Explain why ethanol is described as a biofuel. [2 marks]
   b Ethanol is also described as carbon-neutral. Explain what a carbon-neutral fuel is. [2 marks]

4 Global energy usage is currently about 5.0×10^{20} joules per year. The global population is about 6×10^9 people. Estimate how much energy per second each person uses on average. [4 marks]

Key points

- Your energy demands are met mostly by burning oil, coal, and gas.
- Nuclear power, biofuels, and renewable resources provide energy to generate some of the energy you use.
- Uranium or plutonium is used as the fuel in a nuclear power station. Much more energy is released per kilogram from uranium or plutonium than from fossil fuels.
- Biofuels are renewable sources of energy. Biofuels such as methane and ethanol can be used to generate electricity.

P3.2 Energy from wind and water

Learning objectives

After this topic, you should know:

- what a wind turbine is made up of
- how waves can be used to generate electricity
- the type of power station that uses water running downhill to generate electricity
- how the tides can be used to generate electricity.

Figure 1 *A wind farm is a group of wind turbines*

Go further!

The mass m of wind (air) passing through a wind turbine each second is proportional to the wind speed v. If the wind speed doubles, the mass of wind passing through the wind turbine each second also doubles. As kinetic energy $= \frac{1}{2}mv^2$, the kinetic energy of the wind passing through each second is therefore 2^3 times greater, because m increases by ×2 and v^2 increases by ×4. In other words, the power of the wind is proportional to v^3.

Strong winds can cause lots of damage on a very stormy day. Even when the wind is much weaker, it can still turn a wind turbine. Energy from the wind and other sources such as waves and tides is called renewable energy. That's because such natural sources of energy can never be used up because they are always being replenished (i.e., replaced) by natural processes.

As well as this, no fuel is needed to produce electricity from these natural sources, so they are carbon-free to run.

Wind power

A wind turbine is an electricity generator at the top of a narrow tower. The force of the wind drives the turbine's blades around. This turns a generator. The power generated increases as the wind speed increases. Wind turbines are unreliable because when there is little or no wind they do not generate any electricity.

Wave power

A wave generator uses the waves to make a floating generator move up and down. This motion turns the generator so it generates electricity. A cable between the generator and the shoreline delivers electricity to the grid system.

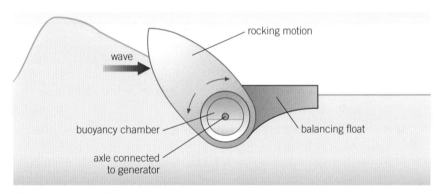

Figure 2 *Energy from waves*

Wave generators need to withstand storms, and they don't produce a constant supply of electricity. Also, lots of cables (and buildings) are needed along the coast to connect the wave generators to the electricity grid. This can spoil areas of coastline. Tidal flow patterns might also change, affecting the habitats of marine life and birds.

Hydroelectric power

Hydroelectricity can be generated when rainwater that's collected in a reservoir (or water in a pumped storage scheme) flows downhill. The flowing water drives turbines that turn electricity generators at the bottom of the hill.

Tidal power

A tidal power station traps water from each high tide behind a barrage. The high tide can then be released into the sea through turbines. The turbines drive generators in the barrage.

One of the most promising sites for a tidal power station in Britain is the Severn estuary. This is because the estuary rapidly becomes narrower as you move up-river away from the open sea. So it funnels the incoming tide and makes it higher.

In some coastal areas, electricity is generated by the tidal flow passing through undersea turbines on the sea bed. Underwater cables are used to connect these turbines to the **national grid**.

1 Hydroelectricity, tidal power, wave power, and wind power are all renewable energy resources.
   a Explain what a renewable energy resource is. [2 marks]
   b Name the renewable energy resource listed above that:
   i does not need energy from the Sun [1 mark]
   ii does not need water and is unreliable. [1 mark]

2 **a** Use the table below for this question. The output of each source is given in millions of watts (MW).
   i Calculate how many wind turbines would give the same total power output as a tidal power station. [1 mark]
   ii Calculate how many kilometres of wave generators would give the same total output as a hydroelectric power station. [1 mark]
   b Use the words below to complete the location column in the table. [2 marks]
   coastline   estuaries   hilly or coastal areas   mountain areas

	Output	Location	Total cost in £ per MW
Hydroelectric power station	500 MW per station		50
Tidal power station	2000 MW per station		300
Wave power generators	20 MW per kilometre of coastline		100
Wind turbines	2 MW per wind turbine		90

3 The last column of the table above shows an estimate of the total cost per MW of generating electricity using different renewable energy resources. The total cost for each resource includes its running costs and the capital costs to set it up.
   a The capital cost per MW of a tidal power station is much higher than that of a hydroelectric power station. Give one reason for this difference. [2 marks]
   b **i** Name the energy resource that has the lowest total cost per MW. [1 mark]
   ii Give two reasons why this resource might be unsuitable in many locations. [2 marks]

4 **a** Explain what a pumped storage scheme is. [2 marks]
   b Describe the main benefit to electricity users of a pumped storage scheme. [2 marks]

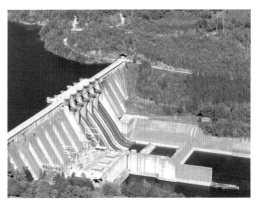

Figure 3 *A hydroelectric power station. Some hydroelectric power stations are designed as pumped storage schemes. When electricity demand is low, electricity can be supplied from other power stations and electricity generators to pumped storage schemes to pump water uphill into a reservoir. When demand is high, the water can be allowed to run downhill to generate electricity*

Figure 4 *A tidal power station*

Key points

- A wind turbine is an electricity generator on top of a tall tower.
- Waves generate electricity by turning a floating generator.
- Hydroelectricity generators are turned by water running downhill.
- A tidal power station traps each high tide and uses it to turn generators.

Learning objectives

After this topic, you should know:

- what solar cells are and how they are used
- the difference between a panel of solar cells and a solar heating panel
- what geothermal energy is
- how geothermal energy can be used to generate electricity.

Solar radiation transfers energy to you from the Sun. That can sometimes be more energy than is healthy if you get sunburnt. But the Sun's energy can be used to generate electricity using solar cells. The Sun's energy can also be used to heat water directly in solar heating panels.

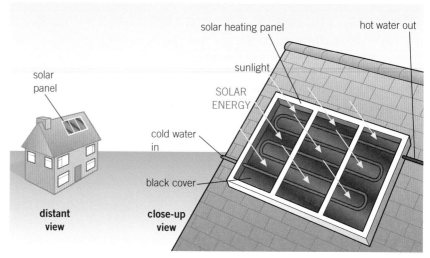

Figure 1 *Solar water heating*

Today's solar cells convert less than 10% of the solar energy they absorb into the energy transferred by electricity. They can be connected together to make solar cell panels.

- They are useful where only small amounts of electricity are needed (e.g., in watches and calculators) or in remote places (e.g. on small islands in the middle of an ocean).
- They are very expensive to buy but they cost nothing to run.
- Lots of them are needed – and plenty of sunshine – to generate enough power to be useful. Solar panels can be unreliable in areas where the Sun is often covered by clouds.

A solar heating panel heats water that flows through it. On a sunny day in Britain, a solar heating panel on a house roof can supply plenty of hot water for domestic use (Figure 1).

A solar power tower uses thousands of flat mirrors to reflect sunlight on to a big water tank at the top of a tower (Figure 3). The mirrors on the ground surround the base of the tower.

- The water in the tank is turned to steam by the heating effect of the solar radiation directed at the water tank.
- The steam is piped down to ground level, where it turns electricity generators.
- The mirrors are controlled by a computer so that they track the Sun.

A solar power tower in a hot dry climate can generate more than 20 MW of electrical power, which is enough to power a few thousand homes.

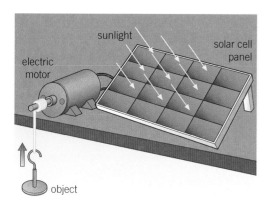

Figure 2 *Solar cells at work*

Figure 3 *A solar power tower*

Geothermal energy

Geothermal energy comes from energy released by radioactive substances deep within the Earth.

- The energy transferred from these radioactive substances heats the surrounding rock.

- So energy is transferred by heating towards the Earth's surface.

Geothermal power stations can be built in volcanic areas or where there are hot rocks deep below the surface. Water gets pumped down to these rocks to produce steam. Then the steam that is produced drives electricity turbines at ground level (Figure 4).

In some areas, buildings can be heated using geothermal energy directly. Heat flow from underground is sometimes called ground source heat. It can be used to heat water in long underground pipes. The hot water is then pumped around the buildings. In some big eco-buildings, this geothermal heat flow is used as under-floor heating.

1   **a** What is the source of geothermal energy ?   [1 mark]
    b Explain why geothermal energy is more reliable than solar energy for heating water.   [1 mark]

2   A satellite in space uses a solar cell panel for electricity. The panel generates 300 W of electrical power and has an area of 10 m^2.
    a Each cell generates 0.2 W. Calculate how many cells are in the panel.   [2 marks]
    b The satellite carries batteries that are charged by electricity from the solar cell panels. Explain why batteries are carried as well as solar cell panels.   [1 mark]

3   A certain geothermal power station has a power output of 200 000 W.
    a Calculate how many kilowatt-hours of energy the power station generates in 24 hours.   [2 marks]
    b Give one advantage and one disadvantage of a geothermal power station compared with a wind turbine.   [2 marks]

4   A solar water panel heats the water flowing through from 14 °C to 35 °C when water flows through it at a rate of 0.010 kilograms per second.
    a Calculate the energy per second transferred by the hot water from the solar panel.   [2 marks]
    b Estimate the output temperature of the hot water if the flow rate is reduced to 0.007 kg/s. The specific heat capacity of water is 4200 J/kg °C   [3 marks]

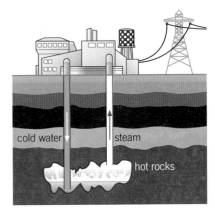

Figure 4 *A geothermal power station*

Key points

- Solar cells are flat solid cells and use the Sun's energy to generate electricity directly.
- Solar heating panels use the Sun's energy to heat water directly.
- Geothermal energy comes from the energy transferred by radioactive substances deep inside the Earth.
- Water pumped into hot rocks underground produces steam to drive turbines at the Earth's surface that generate electricity.

P3.4 Energy and the environment

Learning objectives

After this topic, you should know:

- what fossil fuels do to your environment
- why people are concerned about nuclear power
- the advantages and disadvantages of renewable energy resources
- how to evaluate the use of different energy resources.

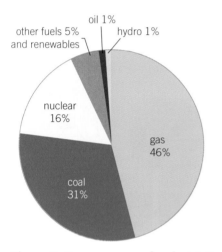

Figure 1 *Energy sources for electricity*

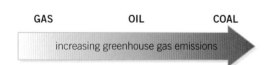

Figure 2 *Greenhouse gases from fossil fuels*

Can you get energy without creating any problems? Figure 1 shows the energy sources people use today to generate electricity. What effect does each one have on your environment?

Fossil fuel problems

When coal, oil, or gas is burnt, greenhouse gases such as carbon dioxide are released. The amount of these gases in the atmosphere is increasing, and most scientists believe that this is causing more global warming and climate change. Some electricity comes from oil-fired power stations. People use much more oil to produce fuels for transport.

Burning fossil fuels can also produce sulfur dioxide. This gas causes acid rain. The sulfur can be removed from a fuel before burning it, to stop acid rain. For example, natural gas has its sulfur impurities removed before it is used.

Fossil fuels are non-renewable. Sooner or later, people will have used up the Earth's reserves of fossil fuels. Alternative sources of energy will then have to be found. But how soon? Oil and gas reserves could be used up within the next 50 years. Coal reserves will last much longer.

Carbon capture and storage (CCS) technology could be used to stop carbon dioxide emissions into the atmosphere from fossil fuel power stations. Old oil and gas fields could be used for carbon dioxide storage.

Nuclear versus renewable

People need to use fewer fossil fuels in order to stop global warming. Should people rely on nuclear power or on renewable energy in the future?

Nuclear power

Advantages

- No greenhouse gases (unlike fossil fuel).
- Much more energy is transferred from each kilogram of uranium (or plutonium) fuel than from fossil fuel.

Disadvantages

- Used fuel rods contain radioactive waste, which has to be stored safely for centuries.
- Nuclear reactors are safe in normal operation. However, an explosion in a reactor could release radioactive material over a wide area. This would affect this area, and the people living there, for many years.

Renewable energy sources and the environment

Advantages

- They will never run out because they are always being replenished by natural processes.
- They do not produce greenhouse gases or acid rain.
- They do not create radioactive waste products.

- They can be used where connection to the National Grid is uneconomical. For example, solar cells can be used for road signs and to provide people with electricity in remote areas.

Disadvantages

- Renewable energy resources are not currently able to meet the world demand. So fossil fuels are still needed to provide some of the energy demand.

- Wind turbines create a whining noise that can upset people nearby, and some people consider them unsightly.

- Tidal barrages affect river estuaries and the habitats of creatures and plants there.

- Hydroelectric schemes need large reservoirs of water, which can affect nearby plant and animal life. Habitats are often flooded to create dams.

- Solar cells need to cover large areas to generate large amounts of power.

- Some renewable energy resources are not available all the time or can be unreliable. For example, solar power is not produced at night and is affected by cloudy weather. Wind power is reduced when there is little or no wind, and hydroelectricity is affected by droughts if reservoirs dry up.

Figure 3 *The effects of acid rain*

1 **a** Name the type of fuel that is used to generate most of Britain's electricity. [1 mark]
   b Give two problems caused by burning fossil fuel. [2 marks]
   c Give two advantages and two disadvantages of using renewable energy sources instead of fossil fuels. [4 marks]

2 Match each energy source with a problem it causes:

Energy source	Problem	
a Coal	**A** Noise	
b Hydroelectricity	**B** Acid rain	
c Uranium	**C** Radioactive waste	
d Wind power	**D** Takes up land	[2 marks]

3 **a** Name three possible renewable energy resources that could be used to generate electricity for people on a remote flat island in a hot climate. [3 marks]
   b Name three types of power stations that do not release greenhouse gases into the atmosphere. [3 marks]

4 A tidal power station, a nuclear power station, or 1000 wind turbines can each supply enough power to meet the electricity needs of a large city on an estuary. Describe the advantages and disadvantages of each type of power station for this purpose. 🖉 [5 marks]

P3.5 Big energy issues

Learning objectives

After this topic, you should know:

- how best to use electricity supplies to meet variations in demand
- how the economic costs of different energy resources compare
- which energy resources need to be developed to meet people's energy needs in future.

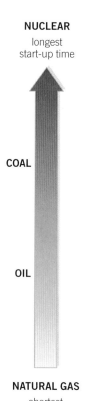

Figure 2 *Start-up times of different types of power station*

Supply and demand

The demand for electricity varies during each day. It is also higher in winter than in summer. Electricity generators need to match these changes in demand.

Power stations can't just start up instantly. The start-up time depends on the type of power station.

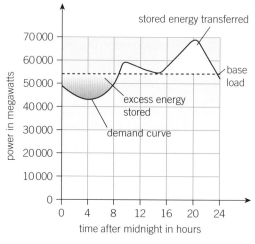

Figure 1 *Example of electricity demand*

Renewable energy resources are unreliable. The amount of electricity they generate depends on the conditions.

Table 1 *Reliability problems with renewable energy resources*

Hydroelectric	Upland reservoir could run dry
Wind, waves	Wind and waves too weak on very calm days
Tidal	Height of tide varies both on a monthly and yearly cycle
Solar	No solar energy at night, and variable during the day

The variable demand for electricity is met by:

- using nuclear and coal-fired power stations to provide a constant amount of electricity (the base load demand)
- using gas-fired power stations and pumped-storage schemes to meet daily variations in demand and extra demand in winter
- using renewable energy resources when demand is high and when the conditions for renewable energy generation are suitable (e.g., use of wind turbines in winter and when wind speeds are high enough)
- using renewable energy resources when demand is low to store energy in pumped storage schemes.

Cost comparisons

The overall cost of a new energy facility involves capital costs to build it, running costs for fuel and maintenance, and more capital costs to take it out of use at the end of its working lifetime.

You can use Table 2 to compare the capital and overall costs of each resource with the associated costs of gas. In practice, capital costs are estimated in terms of cost per kilowatt of power, but running costs are estimated in terms of cost per kilowatt hour of energy.

Table 2 *Cost comparisons (2012). ᵃcoal/gas with carbon capture storage(CCS); ᵇnuclear (excluding decommissioning costs); ᶜoffshore wind; ᵈsolar cell panels.*

Resource	gasᵃ	coalᵃ	uraniumᵇ	hydroelectric	windᶜ	solarᵈ
Capital costs	1.0–2.0	5.5	5.0	3.0	6.0	4.0
Overall costs	1.0–1.2	1.5	1.1	0.7	2.2	1.9

Key points from Table 2 include:

- Capital costs are lowest for gas-fired power stations and greatest for wind power and nuclear power, including decommissioning costs (i.e., taking stations out of use).

- Overall costs including fuel costs are the lowest for hydroelectricity, and greatest for offshore wind farms.

The costs of new energy facilities are usually passed on to consumers through increased fuel bills. Energy-saving schemes such as low-energy light bulbs in your home would reduce the need for more power stations. Schemes such as improved home insulation reduce the demand for non-renewable energy resources (e.g., gas). Home owners pay upfront and eventually get their money back through reduced fuel bills.

1 a Name the type of power station that can be started fastest. [1 mark]
   b Name the type of power station that does not produce
      greenhouse gases or radioactive waste. [1 mark]
   c Name the types of renewable resource that are unreliable. [1 mark]
   d Name the type of renewable resource that can be used to store
      energy at times of low electricity demand. [1 mark]

2 People need to cut back on fossil fuels to reduce the production of
   greenhouse gases. What could happen if the only energy people
   used was:
   a renewable energy [1 mark] b nuclear power? [1 mark]

3 Explain why nuclear power stations are unsuitable for meeting daily
   variations in the demand for electricity. 🚫 [1 mark]

4 Explain what pumped storage schemes are and why they
   are useful. [3 marks]

5 Using the data in Table 2, compare the use of fossil fuel resources
   with renewable resources.
   a Name the resource that has the lowest capital costs. [1 mark]
   b Hydroelectricity has the lowest overall costs. Give a
      reason why its overall cost is less than:
      i   wind or solar power [1 mark]
      ii the non-renewable resources listed in Table 2. [2 marks]
   c The overall cost of a nuclear power station is about 9 pence
      per kW h. Nuclear reactor decommissioning costs have been
      estimated by the UK government at £1000 million per reactor.
      Estimate the extra overall cost per kW h of decommissioning a
      4000 MW reactor that has a lifetime of 30 years. [3 marks]

Key points

- Gas-fired power stations and pumped-storage stations can meet variations in demand.
- Nuclear power stations are expensive to build, run, and decommission. Carbon capture of fossil fuel emissions is likely to be very expensive. Renewable resources are cheap to run but expensive to install.
- Nuclear power stations, fossil-fuel power stations that use carbon capture technology, and renewable energy resources are all likely to be needed for future energy supplies.

P3 Energy resources

Summary questions

1 a i Explain what is meant by a renewable energy source. [2 marks]

ii Explain with the aid of a suitable example what is meant by renewable fuel. [2 marks]

b Evaluate whether or not renewable fuels contribute to greenhouse gases in the atmosphere. [2 marks]

2 a Compare a tidal power station and a hydroelectric power station. Give two similarities and two differences. [4 marks]

b Name the renewable energy resource that transfers:

i the kinetic energy of moving air to energy transferred by an electric current [1 mark]

ii the gravitational potential energy of water running downhill into energy transferred by an electric current [1 mark]

iii the kinetic energy of water moving up and down to energy transferred by an electric current. [1 mark]

3 a i Name the energy resource that does not produce greenhouse gases and uses energy which is from inside the Earth. [1 mark]

ii Name the energy resource that uses running water and does not produce greenhouse gases. [1 mark]

iii Name the energy resource that releases greenhouse gases and causes acid rain. [1 mark]

iv Name the energy resource that does not release greenhouse gases but does produce waste products that need to be stored for many years. [1 mark]

b Wood can be used as a fuel. Determine whether it is

i renewable or non-renewable [1 mark]

ii a fossil fuel or a non-fossil fuel. [1 mark]

4 a Figure 1 shows a landscape showing three different renewable energy resources, numbered 1 to 3.

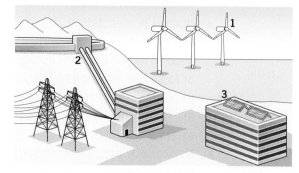

Figure 1 *Renewable energy*

Match each type of energy resource with one of the labels below.

hydroelectricity solar energy wind energy [2 marks]

b Determine which of the three resources shown is not likely to produce as much energy as the others if the area is:

i hot, dry, and windy [1 mark]

ii wet and windy. [1 mark]

5 a Use the data in Table 2 in Topic P3.5 to evaluate whether or not wind turbines are less expensive to build than nuclear power stations, and if they are also cheaper to run. ✏ [4 marks]

b Evaluate the reliability and environmental effects of the non-fossil fuel resources listed in Table 2 in Topic P3.5. ✏ [6 marks]

6 A hydroelectric power station has an upland reservoir that is 400 m above the power station. The power station is designed to produce 96 MW of electrical power with an efficiency of 60%.

a Estimate the loss of gravitational potential energy per second when the hydroelectric power station generates 96 kW of power. [2 marks]

b Use your estimate to calculate the volume of water per second that flows from the reservoir through the power station generators when 96 kW of power is generated. The density of water is $1000 \, kg/m^3$. [3 marks]

7 The national demand for electricity varies during the day and also from winter to summer. Different types of power stations are used to meet these variations in demand. The lowest level of demand for power (the base load demand) is during the daytime in summer.

a Give one reason why the demand for electricity is lower in summer than in winter. [1 mark]

b Different types of power stations are listed below.

coal	oil	gas
nuclear	geothermal	hydroelectric

i Explain which two types of power station from the options above should be connected to the grid system to supply extra power if the demand suddenly increases. [2 marks]

ii Which type of power station is the least suitable for connecting to the grid to meet a sudden increase in demand? Give a reason for your answer. [3 marks]

c Evaluate whether wind turbines or solar panels could be used to meet a sudden increase in demand for energy. [4 marks]

Practice questions

01.1 Which of the following statements are good reasons for using renewable energy resources?

    A supplies of renewable energy are unlimited

    B renewable energy cannot generate electricity all the time

    C renewable energy can be replenished

    D renewable energy cannot supply electricity to millions of homes

    E renewable energy does not produce carbon dioxide   [2 marks]

02.1 **Figure 1** shows a lagoon tidal barrage system used to generate electricity.

Figure 1

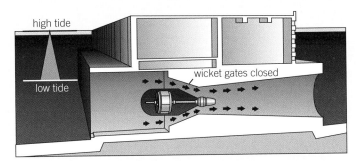

The sentences are not in the correct order. Write sentences **A** to **E** in the correct order, with **A** first.

    A When the tide is high the gates open and water rushes through the turbines into the lagoon.

    B The water flowing out again turns the turbines and generates electricity.

    C The water coming in turns the turbines and generates electricity.

    D When the tide is low the gates are opened and the water rushes out of the lagoon.

    E The lagoon fills up with water and the gates are closed.   [4 marks]

02.2 The lagoon tidal barrage system is claimed to be able to generate electricity for 14 hours a day. Give two reasons why it is necessary to connect houses to the National Grid rather than to the lagoon tidal barrage system alone.   [2 marks]

02.3 Suggest a reason, apart from cost, why people may object to the planned system.   [1 mark]

03 A group of students investigated whether changing the direction of a solar panel affected the amount of electricity generated. The tests were performed on a sunny day at midday. The students method is given below.

1 Fix the solar panel to a board and incline the panel 45° to the horizontal.

2 Attach a voltmeter to the solar panel.

3 Point the solar panel to the North and record the voltage reading.

4 Repeat the test pointing the solar panel in different directions.

The students recorded their results in **Table 1**.

Table 1

Direction	N	NE	E	SE	S	SW	W	NW
Volts	1.2	1.9	2.5	3.9	4.3	3.7	2.4	1.8

03.1 Name the independent and dependent variable.   [2 marks]

03.2 Suggest what advice you would give to a school about to install solar panels.   [1 mark]

03.3 One student stated that the voltage readings would be higher if the investigation was carried out on the roof. Is the student correct? Give a reason for your decision.   [2 marks]

03.4 The students want to check if their results are repeatable. They intend to carry out the investigation using the same apparatus at the same location. Name one other factor they must keep the same.   [1 mark]

04.1 The UK Government has set a target to use renewable energy for 20% of the country's energy use by 2020. Explain why it is important to increase the use of renewable energy sources to generate electricity.   [3 marks]

04.2 The UK Government has agreed to build a nuclear power plant by 2025. It is estimated that the power plant will provide at least 7% of the electricity needed in the UK.

Give two advantages and one disadvantage of using nuclear energy to produce electricity.   [3 marks]

05 Electricity can be generated using ethanol as an energy source. Ethanol can be produced from sugar cane. Brazil is the biggest grower of sugar cane, and the biggest exporter of ethanol made from sugar cane in the world.

Evaluate the benefits of growing and using ethanol made from sugar cane. Your answer should include environmental as well as economic reasons.   [3 marks]

2 Particles at work

All substances are made of atoms. Most atoms are stable and remain stable. Without this, the world as we know it wouldn't exist, and neither would we.

Every atom contains a nucleus surrounded by tiny particles called electrons. Atoms can lose or gain electrons, with different results. For example:

- Materials have different properties when the electrons in their atoms are shared in different ways.

- Metals conduct electricity because they contain electrons that have broken away from atoms inside the metal.

- Radioactive substances are made of atoms with unstable nuclei that emit harmful radiation when they become stable.

In this section you will learn about atoms as you learn about materials, electricity, and radioactivity.

Key questions

- What is an electric current?

- How do series and parallel circuits differ?

- What do we mean by density and elasticity?

- What is the half-life of a radioactive isotope?

Making connections

- The electricity you use at home is produced by generators in power stations and is used by electric motors in appliances such as washing machines. You will learn how motors work in **P13 Electromagnetism**.

- Strong lightweight materials are used for many purposes, including safety equipment such as cycle helmets. Density and elasticity are important properties of such materials. You will meet the use of these materials in **P10 Force and motion**.

I already know...

There are two types of electric charge.

Potential difference is measured in volts and current is measured in amperes.

A cell or battery pushes electrons round a circuit.

Power is how much energy is transferred per second.

Mass is the amount of matter in a substance and is measured in kilograms.

Gas particles move about very quickly and collide with the surface of the gas container.

The nucleus of an atom is composed of protons and neutrons.

I will learn...

How to calculate the charge flow in an electric circuit.

How to work out the resistance and potential difference in an electric circuit.

How mains electricity differs from electricity supplied by batteries.

How to calculate the power of an electrical appliance.

What we mean by density and how we can measure it.

How to explain why the pressure of a gas increases when it is heated in a sealed container.

How an unstable nucleus changes when it becomes stable and why the radiation it gives out is harmful.

Required Practicals

Practical		Topic
15	Investigating resistance	P4.2 P4.5
16	Investigating electrical components	P4.3
17	Measuring the density of a solid object and of a liquid	P6.1

Learning objectives

After this topic, you should know:

- how electric circuits are shown as diagrams
- the difference between a battery and a cell
- what determines the size of an electric current
- how to calculate the size of an electric current from the charge flow and the time taken.

If you experience a power cut at night, an electric torch can be very useful. But it needs to be checked regularly to make sure it works. Figure 1 shows what is inside a torch. The circuit shows how the torch bulb is connected to the switch and the two cells.

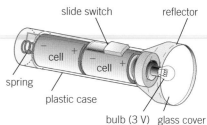

Figure 1 *An electric torch*

A circuit diagram shows you how the components in a circuit are connected together. Each component has its own symbol. Figure 2 shows the symbols for some of the components you will meet in this course. The function of each component is also described. You need to recognise these symbols and remember what each component is used for – otherwise you will get mixed up. More importantly, if you mix them up when building a circuit you could get a big shock.

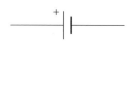 A cell is necessary to push electrons around a complete circuit. A battery consists of two or more cells. The + symbol next to the long line of the cell indicates that this is the positive terminal of the cell.

A switch enables the current in a circuit to be switched on or off.

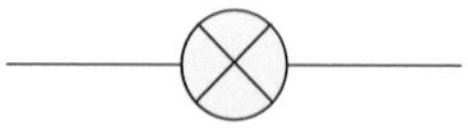

An indicator, such as a bulb, is designed to emit light as a signal when a current passes through it.

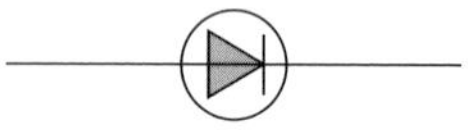

A diode allows current through in one direction only.

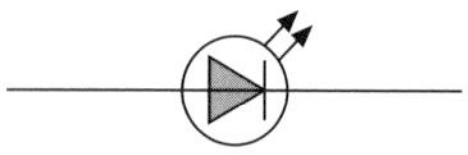

A light-emitting diode (LED) emits light when a current passes through it.

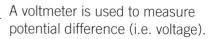 A fixed resistor limits the current in a circuit.

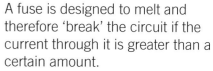 A variable resistor allows the current to be varied.

 A fuse is designed to melt and therefore 'break' the circuit if the current through it is greater than a certain amount.

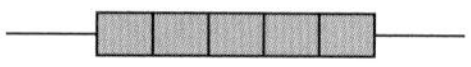

A heater is designed to transfer the energy from an electric current to heat the surroundings.

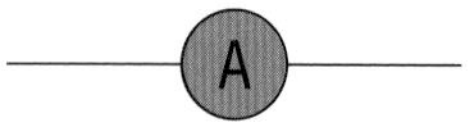

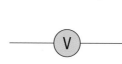  An ammeter is used to measure electric current.

A voltmeter is used to measure potential difference (i.e. voltage).

Figure 2 *Components and symbols*

Electric current

An electric current is a flow of charge. When an electric torch is on, millions of **electrons** pass through the torch bulb and through the cell every second (Figure 3). Each electron carries a negative charge. Metals contain lots of electrons that move about freely between the positively charged metal ions. These electrons stop the ions moving away from each other. The electrons pass through the bulb because it is in a circuit and the bulb filament is metal. The current in the circuit transfers energy from the cell to the torch bulb.

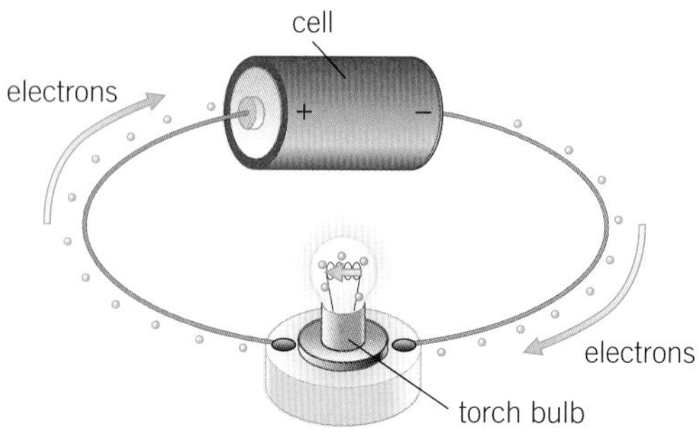

The size of an electric current is the rate of flow of electric charge. This is the flow of charge per second. The bigger the number of electrons that pass through a component each second, the bigger is the current passing through it.

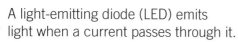

Figure 3 *Electrons on the move*

In a circuit that is a single closed loop like in Figure 3, the current at any point in the circuit is the same as the current at any other point in the circuit. This is because the number of electrons per second that pass through any part of the circuit is the same as at any other part.

Electric charge is measured in coulombs (C). Electric current is measured in amperes (A), sometimes abbreviated as 'amps'.

An electric current of 1 ampere is a rate of flow of charge of 1 coulomb per second. When there is a constant current in a wire (or component) in a circuit for a given time,

charge flow, Q   =   **current, I**   ×   **time taken, t**

(coulombs, C)   (amperes, A)   (seconds, s)

Circuit tests

Connect a variable resistor in series with the torch bulb and a cell (Figure 4). Adjust the slider of the variable resistor. This alters the amount of current flowing through the bulb and so affects its brightness.

- In Figure 4, the torch bulb goes dim when the slider is moved one way. Describe what happens if the slider is moved back again.
- Describe what happens if you include a diode in the circuit.

More about diodes

You would damage a portable radio if you put the batteries in the wrong way around, unless a diode were in series with the battery. The diode allows current through only when it is connected as shown in Figure 5.

1 Name the numbered components in the circuit diagram in Figure 6.

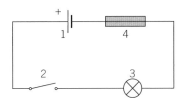

Figure 6   [2 marks]

2 a Redraw the circuit diagram in Figure 6 with a diode in place of the switch so it allows current through. [1 mark]
   b Name the further component you would need in this circuit to change the current in it. [1 mark]
   c When the switch is closed in Figure 6, a current of 0.25 A passes through the lamp. Calculate the charge that passes through the lamp in 60 seconds. [2 marks]

3 a Explain what an ammeter is used for. [1 mark]
   b Explain what a variable resistor is used for. [1 mark]

4 a Draw a circuit diagram for the electric torch in Figure 1. [2 marks]
   b Describe the energy transfers of an electron around the circuit when the switch is closed. ✐ [3 marks]

Worked example

A charge of 16.0 C passes through a bulb in 5.0 seconds. Calculate the current through the bulb.

Solution

$$I = \frac{Q}{t} = \frac{16.0\,C}{5.0\,s} = 3.2\,A$$

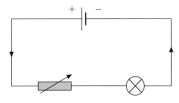

Figure 4 *Using a variable resistor*

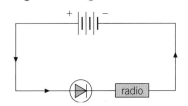

Figure 5 *Using a diode*

Key points

- Every component has its own agreed symbol. A circuit diagram shows how components are connected together.
- A battery consists of two or more cells connected together.
- The size of an electric current is the rate of flow of charge.
- The equation for the electric current of a circuit is $I = \dfrac{Q}{t}$
- The above equation can be rearranged to find charge flow or time: $Q = I\,t$ or $t = \dfrac{Q}{I}$

P4.2 Potential difference and resistance

Learning objectives

After this topic, you should know:

- what is meant by potential difference
- what resistance is and what its unit is
- what Ohm's law is
- what happens when you reverse the potential difference across a resistor.

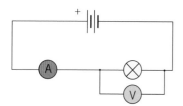

Figure 1 *Using an ammeter and a voltmeter*

Worked example

The energy transferred to a bulb is 320 J when 64 C of charge passes through it. Calculate the potential difference across the bulb.

Solution

$$V = \frac{E}{Q} = \frac{320\,J}{64\,C} = \textbf{50 V}$$

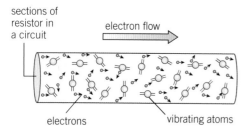

Figure 2 *Electrons moving through a resistor*

Rearranging equations

Rearranging the equation $R = \dfrac{V}{I}$ gives:

$$V = I \times R \quad \text{or} \quad I = \frac{V}{R}$$

Potential difference

Look at the circuit in Figure 1. The battery forces electrons to pass through the ammeter and the bulb.

- The ammeter measures the current through the torch bulb. It is connected in **series** with the bulb, so the current through them is the same. The ammeter reading tells you the current in amperes (or milliamperes (mA) for small currents, where 1 mA = 0.001 A).

- The voltmeter measures the **potential difference** (p.d. or voltage) across the torch bulb. This is the energy transferred to the bulb or the work done on it by each coulomb of charge that passes through it. The unit of potential difference is the volt (V).

- The voltmeter is connected in **parallel** with the torch bulb, so it measures the potential difference across it. The voltmeter reading tells you the potential difference in volts (V).

When charge flows steadily through an electronic component:

$$\textbf{potential difference across a component, } V \textbf{ (volts, V)} = \frac{\textbf{energy transferred, } E \textbf{ (joules, J)}}{\textbf{charge, } Q \textbf{ (coulombs, C)}}$$

Resistance

Electrons passing through a torch bulb have to push their way through lots of vibrating atoms in the metal filament. The atoms resist the passage of electrons through the torch bulb.

The **resistance** of an electrical component is defined as:

$$\textbf{resistance, } R \textbf{ (ohms, } \Omega\textbf{)} = \frac{\textbf{potential difference, } V \textbf{ (volts, V)}}{\textbf{current, } I \textbf{ (amperes, A)}}$$

The unit of resistance is the ohm (Ω). Note that a resistor in a circuit limits the current. For a given potential difference, the larger the resistance of a resistor, the smaller the current.

Current–potential difference graphs

Look at Figure 3. Part **a** shows a circuit that can be used to investigate how the current in a wire depends on the potential difference across the wire. The graph in part **b** shows the measurements and is a straight line through the origin. This means that the current is directly proportional to the potential difference. In other words, the resistance (= potential difference ÷ current of the wire) is constant. This was first discovered for a wire at constant temperature by Georg Ohm and is called **Ohm's law:**

The current through a resistor at constant temperature is directly proportional to the potential difference across the resistor.

A wire is called an ohmic conductor because its resistance stays constant as the current changes, provided its temperature is constant. Figure 4 shows that, reversing the potential difference (and therefore the current) makes no difference to the shape of the line. The resistance is the same whichever direction the current is in.

The gradient of the line depends on the resistance of the resistor. The greater the resistance of the resistor, the less steep the line.

How does the resistance of a wire depend on its length?
Use the circuit in Figure 3 to measure the p.d. for the same current across different lengths of wire.

● Plot a graph of resistance on the *y*-axis against length on the *x*-axis.

● Use your graph to draw conclusions about how resistance varies with length.

Safety: Make sure the wire does not get hot. If it does get hot, reduce the current or switch the circuit off, and ask your teacher to check it.

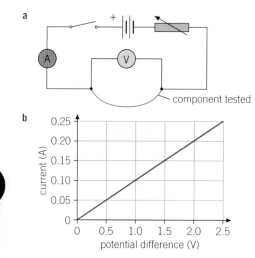

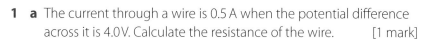

Figure 3 *Investigating the resistance of a wire* **a** *Circuit diagram* **b** *A current– potential difference graph for a wire*

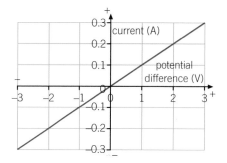

Figure 4 *A current–potential difference graph for a resistor*

1 a The current through a wire is 0.5 A when the potential difference across it is 4.0 V. Calculate the resistance of the wire. [1 mark]
  b Calculate the resistance of the wire that gave the results in the graph in Figure 3. [2 marks]
2 Calculate the missing value in each line of this table.

Resistor	Current in amperes	Potential difference in volts	Resistance in ohms
W	2.0	12.0	
X	4.0		20
Y		6.0	3.0

[3 marks]

3 A torch bulb lights normally when the current through it is 0.015 A and the potential difference across it is 12.0 V.
  a Calculate the resistance of the torch bulb when it lights normally. [1 mark]
  b When the torch bulb lights normally, calculate:
    i the charge passing through the torch bulb in 1200 s [1 mark]
    ii the energy delivered to the torch bulb in this time. [1 mark]
4 a Calculate the resistance of the wire that gave the results shown in Figure 4. [2 marks]
  b i Calculate the current through the wire when the p.d. across it is 1.6 V. [2 marks]
    ii Calculate the p.d. across the wire when the current through it is 0.42 A. [2 marks]

Key points

● potential difference across a component, $V = \dfrac{\text{energy transferred } E,}{\text{charge, } Q}$

● resistance, $R = \dfrac{\text{potential difference, } V}{\text{current, } I}$

● Ohm's law states that the current through a resistor at constant temperature is directly proportional to the potential difference across the resistor.

● Reversing the potential difference across a resistor reverses the current through it.

P4.3 Component characteristics

Learning objectives

After this topic, you should know:

- what happens to the resistance of a filament lamp as its temperature increases
- how the current through a diode depends on the potential difference across it
- what happens to the resistance of:
  - a temperature-dependent resistor as its temperature increases
  - a light-dependent resistor as the light level increases.

Have you ever switched a light bulb on only to hear it pop and fail? Electrical appliances can fail at very inconvenient times. Most electrical failures happen because too much current passes through a component in the appliance.

Investigating different components

The resistance of a wire is independent of the current passing through it. You can use the circuit in Figure 1 to find out if the resistance of other components in the circuit depends on the current. You can also see if reversing the component in the circuit has any effect.

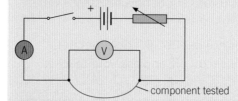

component tested

Figure 1

Make your own measurements using a resistor, a filament lamp, and a diode.

Plot your measurements on a current–potential difference graph. Plot the reverse measurements on the negative section of each axis.

Safety: Make sure the wire does not get hot. If it does get hot, reduce the current or switch the circuit off, and ask your teacher to check it.

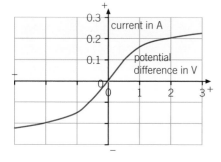

Figure 2 *A current–potential difference graph for a filament lamp*

Using current–potential difference graphs

A filament lamp

Figure 2 shows the graph for a torch bulb (i.e., a low-voltage filament lamp).

- The line curves away from the *y*-axis. So, the current is *not* directly proportional to the potential difference. The filament lamp is a non-ohmic conductor.

- The equation for the resistance of an appliance is:

$$\text{resistance, } R \text{ (ohms, } \Omega) = \frac{\text{potential difference, } V \text{ (volts, V)}}{\text{current, } I \text{ (amperes, A)}}$$

- The resistance increases as the current increases. So, the resistance of a filament lamp increases as the filament temperature increases. The atoms in the metal filament vibrate more as the temperature increases. So they resist the passage of the electrons through the filament more. The resistance of any metal increases as its temperature increases.

- Reversing the potential difference reverses the current and makes no difference to the shape of the curve. The resistance is the same for the same current, regardless of its direction.

Study tip

Try to learn the current–potential difference graphs for a resistor, a filament lamp, and a diode. The filament lamp graph and diode graph are not straight lines.

When a filament light bulb fails, it usually happens when you switch it on. Because resistance is low when the bulb is off, a big current passes through it when you switch it on. If the current is too big, it burns the filament out.

The diode

Figure 3 shows a current–potential difference graph for a diode. The current through a diode flows in one direction only, called the forward direction.

- In the forward direction, the line curves towards the *y*-axis. So the current is not directly proportional to the potential difference. The resistance changes as the current changes. A **diode** is a non-ohmic conductor.

- In the reverse direction, the current is virtually zero. So the diode's resistance in the reverse direction is a lot higher than its resistance is in the forward direction.

A **light-emitting diode (LED)** is a diode that emits light when a current passes through it in the forward direction. LEDs are used as indicators in many electronic devices such as battery chargers and alarm circuits.

Thermistors and light-dependent resistors

Thermistors and light-dependent resistors are used in sensor circuits.

A **thermistor** is a temperature-dependent resistor, and its resistance *decreases* if its temperature increases (and increases if the temperature decreases).

The resistance of a **light-dependent resistor (LDR)** *decreases* if the light intensity increases (and increases if the light intensity decreases).

1 **a** Identify the type of component that has a resistance that:
   i decreases as its temperature increases [1 mark]
   ii depends on which way around it is connected
       in a circuit [1 mark]
   iii increases as the current through it increases. [1 mark]
   b Calculate the resistance of the filament lamp that gave
       the graph in Figure 2 at:
   i 0.1 A **ii** 0.2 A. [2 marks]

2 A thermistor is connected in series with an ammeter and a 9.0 V battery (Figure 5).

   a At 15 °C, the current through the thermistor is 0.6 A, and the potential difference across it is 9.0 V. Calculate its resistance at this temperature. [1 mark]
   b Name and explain what happens to the ammeter reading if the thermistor's temperature is increased. [3 marks]

3 The thermistor in the Figure 5 is replaced by a light-dependent resistor (LDR). Name and explain what happens to the ammeter reading when the LDR is covered. [3 marks]

4 For the diode that gave the graph in Figure 3, when the potential difference is increased steadily from zero, describe how:
   a the current changes [2 marks]
   b the resistance changes. [2 marks]

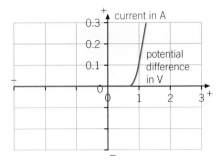

Figure 3 *A current–potential difference graph for a diode*

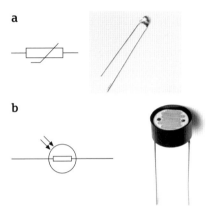

a

b

Figure 4 **a** *A thermistor and its circuit symbol*
b *An LDR and its circuit symbol*

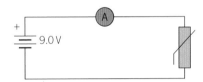

Figure 5 *See Summary question 3*

Key points

- The resistance of an appliance is $R = \dfrac{V}{I}$
- A filament lamp's resistance increases if the filament's temperature increases.
- Diode: forward resistance low; reverse resistance high.
- A thermistor's resistance decreases if its temperature increases.
- An LDR's resistance decreases if the light intensity on it increases.

P4.4 Series circuits

Learning objectives

After this topic, you should know:

- about the current, potential difference, and resistance for each component in a series circuit
- about the potential difference of several cells in series
- how to calculate the total resistance of two resistors in series
- why adding resistors in series increases the total resistance.

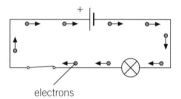

Figure 1 *A torch bulb circuit*

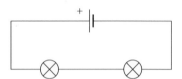

Figure 2 *Bulbs in series*

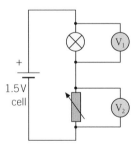

Figure 3 *Voltage tests*

Table 1 *The voltmeter readings for each setting add up to 1.5 V. This is the potential difference of the cell*

Filament lamp	Voltmeter V_1 in volts	Voltmeter V_2 in volts
normal	1.5	0.0
dim	0.9	0.6
very dim	0.5	1.0

Circuit rules

In the torch circuit in Figure 1, the bulb, the cell, and the switch are connected in series with each other. The same number of electrons passes through each component every second. So the same current passes through every component.

In a series circuit, the same current passes through each component.

In Figure 2, each electron from the cell passes through two bulbs. The electrons are pushed through each bulb by the cell. The potential difference (p.d.) of the cell is a measure of the energy transferred from the cell by each electron that passes through it. Because each electron in the circuit in Figure 2 passes through both bulbs, the potential difference of the cell is shared between the bulbs. This rule applies to any series circuit.

In a series circuit, the total potential difference of the power supply is shared between the components.

Cells in series

What happens if you use two or more cells in series in a circuit? As long as you connect the cells so that they act in the same direction, each electron gets a push from each cell. So an electron would get the same push from a battery of three 1.5 V cells in series as it would from a single 4.5 V cell. In other words, as long as the cells act in the same direction:

The total potential difference of cells in series is the sum of the potential difference of each cell.

Investigating potential differences in a series circuit
Figure 3 shows how to test the potential difference rule for a series circuit. The circuit consists of a filament lamp in series with a variable resistor and a cell. You can use the variable resistor to see how the voltmeter readings change when you change the current. Make your own measurements.

- Compare your measurements with the data in Table 1.

Safety: Make sure the wire does not get hot. If it does get hot, reduce the current or switch the circuit off, and ask your teacher to check it.

The resistance rule for components in series

In Figure 3, suppose the current through the bulb is 0.1 A when the bulb is dim. Using data from Table 1:

- the resistance of the bulb would be 9 Ω (= 0.9 V ÷ 0.1 A),
- the resistance of the variable resistor at this setting would be 6 Ω (= 0.6 V ÷ 0.1 A).

If you replaced these two components with a single resistor, its resistance would need to be 15 Ω for the same current of 0.1 A. This is because 1.5 V ÷ 0.1 A is equal to 15 Ω. This resistance is the sum of the resistance of the two components. The same rule applies to any series circuit.

The total resistance of two (or more) components in series is equal to the sum of the resistance of each component.

So for two components of resistances R_1 and R_2 in series (Figure 4):

total resistance, R_{total} $(\Omega) = R_1 + R_2$

R_1 (Ω) is the resistance of the first component and R_2 (Ω) is the resistance of the second component.

Adding more resistors in series increases the total resistance of the circuit. This is because the total potential difference is shared between more resistors, and as a result the potential difference across each of them is less than before. The current through the resistors is therefore less than before, and as the total potential difference is unchanged, the total resistance is therefore greater.

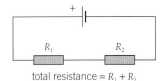

total resistance = $R_1 + R_2$

Figure 4 *Resistors in series*

Worked example

A 4.5 V battery is connected to a 1.0 Ω resistor and a 5.0 Ω in series with each other.

Calculate:

a the total resistance of the two resistors

b the current through the resistors.

Solution

a Total resistance = 1.0 Ω + 5.0 Ω
= 6.0 Ω

b Current = $\dfrac{\text{battery potential difference}}{\text{total resistance}}$

$= \dfrac{4.5\ V}{6.0\ \Omega} = 0.75\ A$

1 **a** In Figure 2, if the potential difference of the cell is 1.2 V and the potential difference across one bulb is 0.8 V, work out the potential difference across the other bulb. [2 marks]

b In Figure 3, the bulb emits light when the resistance of the variable resistor is 5.0 Ω and the potential difference across the variable resistor is 1.0 V. Calculate the current through the bulb and the potential difference across it. [2 marks]

2 A 1.5 V cell is connected to a 3.0 Ω resistor and 2.0 Ω resistor in series with each other.

a Draw the circuit diagram for this arrangement. [1 mark]

b Calculate:

i the total resistance of the two resistors [1 mark]

ii the current through the resistors. [2 marks]

c The 3.0 Ω resistor is replaced by a different resistor X and the current changes to 0.25 A. Calculate the resistance of X. [2 marks]

3 For the circuit in Figure 5, each cell has a potential difference of 1.5 V.

a Calculate:

i the total resistance of the two resistors [1 mark]

ii the total potential difference of the two cells. [1 mark]

b Show that the current through the battery is 0.25 A. [1 mark]

c Calculate the potential difference across each resistor. [2 marks]

d If a 3 Ω resistor R is connected in series with the two resistors, calculate:

i their total resistance [1 mark]

ii the current through the resistors [2 marks]

iii the potential difference across each resistor. [3 marks]

4 Explain why the resistance of several resistors in series is increased if an additional resistor is connected in series with them. 🖊 [4 marks]

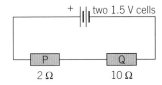

Figure 5 *See Summary question 3*

Key points

- For components in series:
  - the current is the same in each component
  - the total potential difference is shared between the components
  - adding their resistances gives the total resistance.
- For cells in series, acting in the same direction, the total potential difference is the sum of their individual potential differences.
- Total resistance $R_{total} = R_1 + R_2$.
- Adding more resistors in series increases the total resistance because the current through the resistors is reduced and the total potential difference across them is unchanged.

P4.5 Parallel circuits

Learning objectives

After this topic, you should know:

- about the currents and potential differences for components in a parallel circuit
- how to calculate the current through a resistor in a parallel circuit
- why the total resistance of two resistors in parallel is less than the resistance of the smallest individual resistor
- Adding resistors in parallel decreases the total resistance.

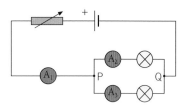

Figure 1 *At a junction*

Table 1

Ammeter readings in amperes		
A_1	A_2	A_3
0.50	0.30	0.20
0.30	0.20	0.10
0.18	0.12	0.06

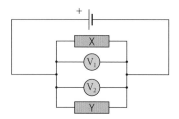

Figure 2 *Components in parallel*

Figure 1 shows how you can investigate the current through two bulbs in parallel with each other. You can use ammeters in series with the bulbs and the cell to measure the current through each component. The bulbs are in separate branches of the circuit.

In each case, the reading of ammeter A_1 is equal to the sum of the readings of ammeters A_2 and A_3. This shows that the current through the cell is equal to sum of the currents through the two bulbs. This rule applies wherever components are in parallel.

The total current through the whole circuit is the sum of the currents through the separate branches.

Potential difference in a parallel circuit

Figure 2 shows two resistors X and Y in parallel with each other. A voltmeter is connected across each resistor. The voltmeter across resistor X shows the same reading as the voltmeter across resistor Y. This is because each electron from the cell either passes through X or passes through Y. So it delivers the same amount of energy from the cell, whichever resistor it goes through. In other words:

For components in parallel, the potential difference across each component is the same.

Parallel routes

You can think of a parallel route as a kind of bypass. A heart bypass is another route for the flow of blood. A road bypass is a road that passes a town centre instead of going through it. In the same way, for components in parallel, charge flows separately through each component. The total flow of charge is the sum of the flow through each component. The flow of charge per second through each component is the current through it. So, for components that are in parallel, the total current is the sum of the currents through each component.

Calculations on parallel circuits

Components in parallel have the same potential difference across them. The current through each component depends on the resistance of the component.

- The bigger the resistance of the component, the smaller the current through it. The component that has the biggest resistance passes the smallest current.

- You can calculate the current passing through each component using the equation:

$$\textbf{current, } I \text{ (amperes, A)} = \frac{\textbf{potential difference, } V \text{ (volts, } V)}{\textbf{component resistance, } R \text{ (ohms, } \Omega)}$$

Rules for resistors in parallel

Adding more resistors in parallel decreases the total resistance. This is because the total potential difference is the same across each resistor. So adding an extra resistor in parallel increases the total current entering the combination. Because the total resistance is equal to the battery potential difference ÷ the total current entering the combination, the total resistance is less than it was before the extra resistor was added.

The total resistance of two (or more) components in parallel is less than the resistance of the resistor with the least resistance.

Testing resistors in series and parallel

Use the circuit in Figure 3 on P4.2 to measure the resistance of two resistors individually, and then in series with each other. The resistance of the resistors in series should equal the sum of their individual resistances.

Then measure the resistance of two resistors individually and then in parallel with each other. Discuss if your results agree with the statement above this practical box.

Safety: Make sure the wire does not get hot. If it does get hot, reduce the current or switch the circuit off, and ask your teacher to check it.

1 a In Table 1, if ammeter A_1 reads 0.40 A and A_2 reads 0.1 A, calculate what A_3 would read. [1 mark]
  b A 3 Ω resistor and a 6 Ω resistor are connected in parallel in a circuit. Identify the resistor that passes the most current. [1 mark]
  c In the circuit shown in Figure 3, calculate what the resistance of a single resistor would be that could replace the three parallel resistors across the 6 V battery and allow the same current to pass through the battery. [2 marks]

2 A 6.0 V battery is connected across a 12 Ω resistor in parallel with a 24 Ω resistor.
  a Draw the circuit diagram for this circuit. [1 mark]
  b i Show that the current through the 12 Ω resistor is 0.50 A [1 mark]
    ii Calculate the current through the 24 Ω resistor. [1 mark]
  c Calculate the current passing through the cell. [2 marks]

3 a In a circuit similar to Figure 3, three resistors $R_1 = 2\,\Omega$, $R_2 = 3\,\Omega$, and $R_3 = 6\,\Omega$ are connected to each other in parallel and to a 6 V battery.
    Draw the circuit diagram and calculate:
    i the current through each resistor [3 marks]
    ii the current through the battery. [1 mark]
  b The 6 Ω resistor in the figure is replaced by a 4 Ω resistor, calculate the new battery current. [2 marks]

4 Explain why the equivalent resistance of the three resistors in the circuit in Question 3 is less than 2 Ω. [4 marks]

Worked example

The circuit diagram in Figure 3 shows three resistors $R_1 = 1\,\Omega$, $R_2 = 2\,\Omega$, and $R_3 = 6\,\Omega$ connected in parallel to a 6 V battery.

Calculate:

a the current through each resistor

b the current through the battery.

Solution

a $I_1 = \dfrac{V_1}{R_1} = \dfrac{6}{1} = \textbf{6 A}$

$I_2 = \dfrac{V_2}{R_2} = \dfrac{6}{2} = \textbf{3 A}$

$I_3 = \dfrac{V_3}{R_3} = \dfrac{6}{6} = \textbf{1 A}$

b The total current from the battery
$= I_1 + I_2 + I_3 = 6\,\text{A} + 3\,\text{A} + 1\,\text{A} = \textbf{10 A}$

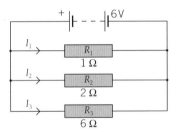

Figure 3

Key points

- For components in parallel:
  - the total current is the sum of the currents through the separate components
  - the potential difference across each component is the same.
- The bigger the resistance of a component, the smaller the current that will pass through that component.
- The current through a resistor in a parallel circuit is $I = \dfrac{V}{R}$
- Adding more resistors in parallel decreases the total resistance because the total current through the resistors is increased and the total potential difference across them is unchanged.

P4 Electric circuits

Summary questions

1 Write and explain how the resistance of a filament lamp changes when the current through the filament is increased. [3 marks]

2 Match each component in the list to each statement **a** to **d** that describes it.

> diode   filament lamp   resistor   thermistor

   a Its resistance increases if the current through it increases. [1 mark]

   b The current through it is proportional to the potential difference across it. [1 mark]

   c Its resistance decreases if its temperature is increased. [1 mark]

   d Its resistance depends on which way around it is connected in a circuit. [1 mark]

3 **a** Sketch a circuit diagram to show two resistors **P** and **Q** connected in series to a battery of two cells in series with each other in the same direction. [1 mark]

   b In the circuit in part **a**, resistor **P** has a resistance of 4.0 Ω, resistor **Q** has a resistance of 6.0 Ω, and each cell has a potential difference of 1.5 V. Calculate:

   i the total potential difference of the two cells [1 mark]

   ii the total resistance of the two resistors [1 mark]

   iii the current in the circuit [2 marks]

   iv the potential difference across each resistor. [2 marks]

4 **a** Sketch a circuit diagram to show two resistors **R** and **S** in parallel with each other connected to a single cell. [1 mark]

   b In the circuit in part **a**, resistor **R** has a resistance of 8.0 Ω, resistor **S** has a resistance of 4.0 Ω, and the cell has a potential difference of 2.0 V. Calculate:

   i the current through resistor **R** [2 marks]

   ii the current through resistor **S** [2 marks]

   iii the current through the cell in the circuit. [1 mark]

5 The figure shows a light-dependent resistor (LDR) in series with a 200 Ω resistor, a 3.0 V battery, and an ammeter.

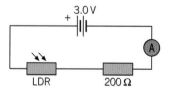

   a With the LDR in daylight, the ammeter reads 0.010 A.

   i Calculate the potential difference across the 200 Ω resistor when the current through it is 0.010 A. [1 mark]

   ii Show that the potential difference across the LDR is 1.0 V when the ammeter reads 0.010 A. [1 mark]

   iii Calculate the resistance of the LDR in daylight. [1 mark]

   b **i** If the LDR is then covered, explain whether the ammeter reading increases or decreases or stays the same. [2 marks]

   ii Explain how the resistance of the LDR can be calculated from the current *I*, the battery potential difference *V*, and the resistance *R* of the LDR. [2 marks]

6 In the figure to Question **5**, the LDR is replaced by a 100 Ω resistor and a voltmeter connected in parallel with this resistor.

   a Draw the circuit diagram for this circuit. [1 mark]

   b Calculate:

   i the total resistance of the two resistors in the circuit [1 mark]

   ii the current through the ammeter [2 marks]

   iii the voltmeter reading [2 marks]

   iv the potential difference across each resistor. [2 marks]

7 The figure shows a light-emitting diode (LED) in series with a resistor and a 3.0 V battery.

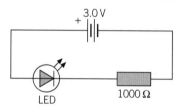

   a The LED in the circuit emits light. The potential difference across it when it emits light is 0.6 V.

   i Explain why the potential difference across the 1000 Ω resistor is 2.4 V. [2 marks]

   ii Calculate the current in the circuit. [2 marks]

   b The current through the LED must not exceed 15 mA or it will be damaged. If the resistor in the figure is replaced by a different resistor **R**, calculate what should be the minimum resistance of **R**. [2 marks]

   c If the LED in the circuit is reversed, what would be the current in the circuit? Give a reason for your answer. [3 marks]

8 **a** Design a temperature sensor that will switch a buzzer on if the temperature is too low. [2 marks]

   b Explain how your circuit works. [3 marks]

Practice questions

01.1 **Figure 1** shows a current–potential difference graph of a fixed resistor.

Figure 1

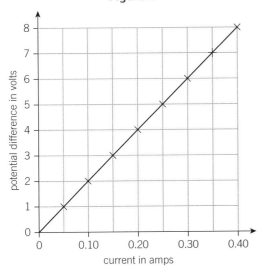

Calculate the resistance of the fixed resistor R_1 and give the unit. [2 marks]

01.2 The fixed resistor R_1 is placed in series with another fixed resistor R_2.

Figure 2

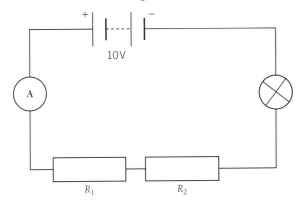

Calculate the total resistance of resistor R_1 and resistor R_2.
Resistance of resistor $R_2 = \frac{2}{3}R_1$ [2 marks]

01.3 A voltmeter is used to measure the potential difference across the bulb. Draw the circuit with the voltmeter in the correct position. [2 marks]

01.4 Calculate the charge and give the unit when a current of 0.15 A flows around the circuit for 20 seconds. [3 marks]

01.5 Calculate how much energy is transferred from the battery in 20 seconds. Use your answer from **01.4**. [2 marks]

02 A student measures the resistance of a coil of wire using the equipment in **Figure 3**.

Figure 3

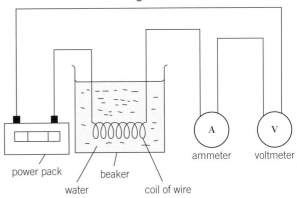

02.1 Give one mistake the student has made in setting up the equipment. [1 mark]

02.2 Give one reason why the coil of wire is in water. [1 mark]

02.3 Describe how the student measures the resistance of the wire using the equipment given. [3 marks]

02.4 Suggest one safety precaution the student should take. [1 mark]

03 A thermistor is used in a circuit to maintain the temperature of a room used to grow tropical plants.

03.1 Draw the circuit symbol for a thermistor. [1 mark]

03.2 Describe what happens to the resistance of the thermistor as the temperature decreases below 20 °C. [2 marks]

03.3 Sketch a graph of current (y-axis) against potential difference (x-axis) for a thermistor that has a constant temperature. [2 marks]

Learning objectives

After this topic, you should know:

- what direct current is and what alternating current is
- what is meant by the live wire and the neutral wire of a mains circuit
- what the National Grid is
- how to use an oscilloscope to measure the frequency and peak potential difference of an alternating current.

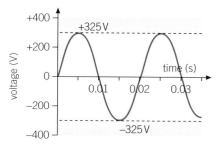

Figure 1 *Mains voltage against time*

The battery in a torch makes the current go around the circuit in one direction only. The current in the circuit is called a **direct current** (d.c.) because it is in one direction only.

When you switch a light on at home, you use **alternating current** (a.c.) because mains electricity is an a.c. supply. An alternating current repeatedly reverses its direction. It flows one way then the opposite way in successive cycles. Its frequency is the number of cycles it passes through each second.

In the UK, the mains frequency is 50 cycles per second (or 50 Hz). The time taken for 1 cycle = 0.02 s (= 1 ÷ frequency). A light bulb works just as well at this frequency as it would with a direct current.

Mains circuits

Every mains circuit has a **live wire** and a **neutral wire**. The current through a mains appliance alternates. That is because the mains supply provides an alternating potential difference between the two wires. In other words, the polarity of the potential difference repeatedly reverses its direction. In comparison, potential differences in direct current circuits do not change direction.

The neutral wire is earthed at the local electricity substation. The potential difference between the live wire and the earth is usually called the potential or voltage of the live wire. The live wire is dangerous because its potential repeatedly changes from + to – and back every cycle. In UK homes, it reaches about 325 V in each direction (Figure 1).

The National Grid

When you use mains appliances, the electricity is supplied from power stations to homes and buildings through the National Grid – a nationwide network of cables and transformers. A typical power station generates electricity at an alternating potential difference of about 25 000 V.

- **Step-up transformers** are used at power stations to transfer electricity to the National Grid. These transformers are used to make the size of the alternating potential difference much bigger, typically from 25 000 V to about 132 000 V).

- **Step-down transformers** are used to supply electricity from the National Grid to consumers. Homes and offices in the UK are supplied with mains electricity that provides the same power as a 230 V direct-current supply. Factories use much more power than homes, so they are supplied with a p.d. of 100 kV or 33 kV.

By making the grid potential difference very large, much less current is needed to transfer the same amount of power. So the power loss due to the resistance heating in the cables is much reduced. This means that the National Grid is an efficient way to transfer power.

Investigating an alternating potential difference

An **oscilloscope** is used to show how an alternating potential difference changes with time.

1 Connect a signal generator to an oscilloscope (Figure 2).

- The waves on the oscilloscope screen are caused by the potential difference increasing and decreasing continuously. Adjusting the 'Y-gain' control changes how tall the waves are. Adjusting the 'time base' control changes how many waves fit across the screen.

- The peak potential difference (peak voltage) is the difference in volts between the highest level and the middle level of the waves. Increasing the potential difference of the a.c. supply makes the waves on the screen taller.

- Increasing the frequency of the a.c. supply increases the number of cycles you see on the screen. So the waves on the screen get squashed together.

2 Connect a battery to the oscilloscope. You should see a flat line at a constant potential difference.

What difference on the oscilloscope screen is made by reversing the battery?

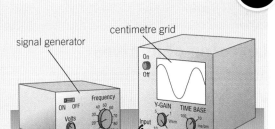

Figure 2 *Using an oscilloscope*

1 Choose the correct potential difference from the list for each appliance **a** to **d**.

1.5V   12V   230V   325V   [4 marks]

a a car battery   **c** a torch cell
b the mains voltage   **d** the maximum potential of the live wire.

2 Describe how the trace on the screen in Figure 2 would change if the frequency of the a.c. supply was:

a increased without changing the peak potential difference [2 marks]
b reduced and the peak potential difference was doubled. [2 marks]

3 Calculate the frequency in Figure 2 if one cycle measures 8 cm across the screen for a time base setting of 10 milliseconds per centimetre.
                                                            [2 marks]

4 **a** Describe how an alternating current differs from a direct current.
                                                            [1 mark]

b Figure 3 shows a diode and a resistor in series with each other and connected to an a.c. supply. Explain why the current in the circuit is a direct current, not an alternating current.   [1 mark]

Figure 3

ac supply

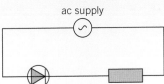

c i Sketch a graph to show how the current varies with time. [1 mark]
ii Suggest how your graph would differ if a resistor of greater resistance had been used.   [2 marks]

Go further!

You'll learn more about alternating current and oscilloscopes at A level, when you carry out experiments such as measuring the frequency of an alternating potential difference.

Key points

- Direct current (d.c.) flows in one direction only. Alternating current (a.c.) repeatedly reverses its direction of flow.
- A mains circuit has a live wire, which is alternately positive and negative every cycle, and a neutral wire at zero volts.
- The peak potential difference of an a.c. supply is the maximum voltage measured from zero volts.
- To measure the frequency of an a.c. supply, measure the time period of the waves, then use the equation

$$\text{frequency} = \frac{1}{\text{time taken for 1 cycle}}$$

P5.2 Cables and plugs

Learning objectives

After this topic, you should know:

- what the casing of a mains plug or socket is made of and why
- what is in a mains cable
- the colours of the live, neutral, and earth wires
- why a three-pin plug includes an earth pin.

When you plug a heater with a metal case into a wall socket, the metal case is automatically connected to earth through a wire called the **earth wire**. This stops the metal case becoming live if the live wire breaks and touches the case. If the case did become live and you touched it, you would be electrocuted.

Plastic materials are good insulators. An appliance that has a plastic case is double-insulated, so it has no earth wire connection. All electrical appliances with a plastic case sold in the UK must be double-insulated.

Plugs, sockets, and cables

The outer casings of **plugs**, sockets, and cables of all mains circuits and appliances are made of hard-wearing electrical insulators. That's because plugs, sockets, and cables contain live wires. Most mains appliances are connected by a wall socket to the mains using a cable and a **three-pin plug**.

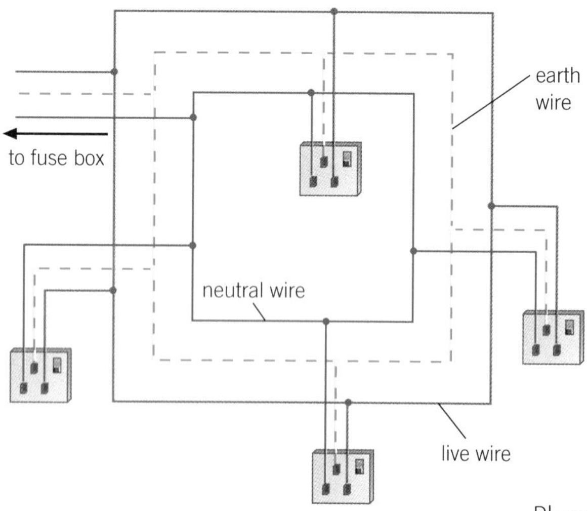

Figure 1 *A wall socket circuit*

Sockets are made of stiff plastic materials with the wires inside them. Figure 1 shows part of a wall socket circuit. It has an earth wire as well as a live wire and a neutral wire.

- The earth wire of this circuit is connected to the ground at your home. It is at zero volts (0 V) and carries a current only if there is a fault.
- The longest pin of a three-pin plug is designed to make contact with the earth wire of a wall socket circuit. So, when you plug an appliance with a metal case into a wall socket, the case is automatically earthed.

Plugs have cases made of stiff plastic materials. The live pin, the neutral pin, and the earth pin stick out through the plug case. Figure 2 shows inside a three-pin plug.

- The pins are made of brass because brass is a good conductor and doesn't rust or oxidise. Copper isn't as hard as brass even though it conducts better.
- The case material is an electrical insulator. The inside of the case is shaped so that the wires and the pins can't touch each other when the plug is sealed.
- The plug contains a **fuse** between the live pin and the live wire. If too much current passes through the wire in the fuse, it melts and cuts the live wire off.
- The brown wire is connected to the live pin.
- The blue wire is connected to the neutral pin.
- The green and yellow striped wire (of a three-core cable) is connected to the earth pin. A two-core cable doesn't have an earth wire.

Study tip

Make sure you know what's inside a three-pin plug and the colour of each wire.

Cables used for mains appliances (and for mains circuits) are made up of two or three insulated copper wires surrounded by an outer layer of rubber or flexible plastic material.

- Copper is used for the wires because it is a good electrical conductor and it bends easily.

- Plastic is a good electrical insulator, so if anyone touches the cable, it stops them from getting an electric shock.

- Two-core cables are used for appliances that have plastic cases (e.g. hairdryers, radios, mobile phone chargers).

- Cables of different thicknesses are used for different purposes. For example, the cables that join together the wall sockets in your house must be much thicker than the cables that join together the light fittings. This is because more current passes along wall socket cables than along lighting circuits, so the wires in them have to be much thicker so they have less resistance. This stops the heating effect of the current making the wires too hot.

Short circuits

If a live wire inside the appliance touches a neutral wire, a very big current passes between the two wires at the point of contact. This is called a short circuit. Provided the fuse blows, it cuts the current off.

Even if an appliance is switched off, never touch the wires inside the supply cable. People's bodies are at zero volts. If someone touches a live wire, a big potential difference will act across their body, causing a current to flow through them. They will suffer an electric shock, which could be lethal.

1   a   Name the colour of each of the three wires in a mains plug.   [1 mark]
   b   i   Explain why sockets are wired in parallel with each other. [2 marks]
      ii   Explain why brass, an alloy of copper and zinc, is better than copper for the pins of a three-pin plug.   [2 marks]
     iii   Give a reason why cables that are worn away or damaged are dangerous.   [1 mark]

2   a   Match the list of parts 1 to 4 in a three-pin plug with the list of materials **A** to **D**.

     1 cable insulation   **A** brass
     2 case   **B** copper
     3 pin   **C** rubber
     4 wire   **D** stiff plastic   [2 marks]
   b   Give your choice of material for each part in **a**.   [4 marks]

3   a   Explain why each of the three wires in a three-core mains cable is insulated.   [2 marks]
   b   Describe how the metal case of an electrical appliance is connected to earth.   [3 marks]

4   a   Explain why the cables joining the wall sockets in a house need to be thicker than the cables joining the light fittings.   [2 marks]
   b   Describe the difference between a two-core cable and a three-core cable.   [1 mark]
   c   Explain what determines whether an appliance should have a two-core or a three-core cable.   [2 marks]

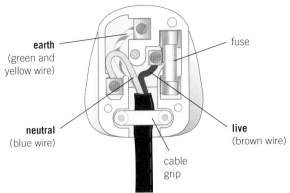

earth (green and yellow wire)   fuse
neutral (blue wire)   **live** (brown wire)
cable grip

Figure 2 *Inside a three-pin plug*

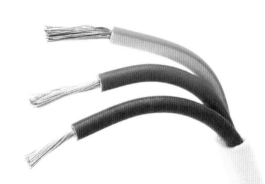

Figure 3 *Mains cables*

Key points

- Sockets and plug cases are made of stiff plastic materials that enclose the electrical connections. Plastic is used because it is a good electrical insulator.
- A mains cable is made up of two or three insulated copper wires surrounded by an outer layer of flexible plastic material.
- In a three-pin plug or a three-core cable, the live wire is brown, the neutral wire is blue, and the earth wire is striped green and yellow.
- The earth wire is connected to the longest pin in a plug and is used to earth the metal case of a mains appliance.

P5.3 Electrical power and potential difference

Learning objectives

After this topic, you should know:

- how power and energy are related
- how to use the power rating of an appliance to calculate the energy transferred in a given time
- how to calculate the electrical power supplied to a device from its current and potential difference
- how to work out the correct fuse to use in an appliance.

Synoptic link

You first met this equation in Topic P1.9.

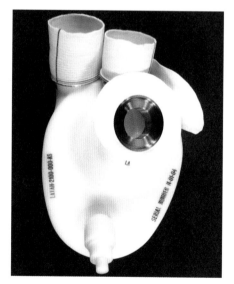

Figure 1 *An artificial heart. A surgeon fitting an artificial heart in a patient needs to make sure the battery will last a long time. Even so, the battery may have to be replaced every few years*

Rearranging equations

Rearranging the equation $P = I V$ gives $V = \dfrac{P}{I}$ or $I = \dfrac{P}{V}$

When you use an electrical appliance, the current through it transfers energy to it from the power source it is connected to. The power of the appliance, in watts, is the energy it transfers in joules per second. You can see this using the equation:

$$\textbf{power, } P \text{ (watts, W)} = \frac{\textbf{energy transferred, } E \text{ (joules, J)}}{\textbf{time, } t \text{ (seconds, s)}}$$

You can calculate the energy transferred E in a given time t if you know the power P, rearrange the above equation to give:

$$E = P\, t$$

Worked example

A 40 W light bulb is switched on for 30 minutes. Calculate the energy it transfers.

Solution

Time taken = $30 \times 60\,\text{s} = 1800\,\text{s}$

$E = P \times t = 40\,\text{W} \times 1800\,\text{s} = 72\,000\,\text{J}$

Calculating power

Millions of electrons pass through the circuit of an artificial heart every second. Work is done by a battery in the artificial heart to force the electrons around the circuit. Each electron transfers a small amount of energy to the heart from the battery. So the total energy transferred to the artificial heart each second is big enough to allow the appliance to function.

For any electrical appliance:

- the current through it is the charge that flows through it each second
- the potential difference across it is the energy transferred to the appliance by each coulomb of charge that passes through it
- the power supplied to it is the energy transferred to it each second. This is the energy transferred by an electric current every second.

So the energy transfer to the appliance each second = the charge flow per second × the energy transfer per unit charge.

In other words:

$$\textbf{power supplied, } P = \textbf{current, } I \times \textbf{potential difference, } V$$
(watts, W)   (amperes, A)   (volts, V)

For example, the power supplied to:

- a 4 A, 12 V electric motor is 48 W (= 4 A × 12 V)
- a 0.1 A, 3 V torch lamp is 0.3 W (= 0.1 A × 3.0 V).

Choosing a fuse

Domestic appliances are often fitted with a 3 A, 5 A, or 13 A fuse. If you do not know which one to use for an appliance, you can calculate it by using the power rating of the appliance and its potential difference (voltage). The next time you change a fuse, do a quick calculation to make sure its rating is correct for the appliance.

Resistance heating

When an electric current passes through a resistor, the power supplied to the resistor heats it. The resistor heats the surroundings, so the energy supplied to it is dissipated to the surroundings.

For a current I in a resistor of resistance R:

- the potential difference V across the resistor $= I \times R$
- the power P supplied to the resistor $= I \times V = I \times I \times R = I^2 R$

power, P (W) = current², I^2 (A) × resistance, R (Ω)

This equation shows that the power supplied to a resistor is proportional to the square of the current. So, for example, if the current is doubled, the power becomes four times greater.

1 a The human heart transfers about 30 000 J of energy in about 8 hours. Estimate to an order of magnitude the power of the human heart. [2 marks]
  b Calculate the power supplied to a 5 A, 230 V electric heater. [2 marks]
  c Explain why a 13 A fuse would be unsuitable for a 230 V, 100 W table lamp. [2 marks]

2 a Calculate the power supplied to each of the following devices in normal use:
  i a 12 V, 5 A light bulb [2 marks]
  ii a 230 V, 12 A heater. [2 marks]
  b Calculate the type of fuse, 3 A, 5 A, or 13 A, that you would select for:
  i a 50 W, 12 V heater [2 marks]
  ii a 230 V, 750 W microwave oven. [2 marks]

3 a Explain why a 3 A fuse would be unsuitable for a 230 V, 800 W microwave oven. [2 marks]
  b The heating element of a 12 V heater has a resistance of 4.0 Ω. When the heating element is connected to a 12 V power supply, calculate:
  i the current through it [2 marks]
  ii the electrical power supplied to it [2 marks]
  iii the energy, in joules, transferred to the heating element by an electric current in 20 minutes. [2 marks]

4 A 6.0 kW electric oven is connected to a fuse box by a cable of resistance 0.25 Ω. When the cooker is switched on at full power, a current of 26 A passes through it.
  a i Calculate the potential difference between the two ends of the cable. [2 marks]
  ii Calculate the power wasted in the cable because of the heating effect of the current. [2 marks]
  b Calculate what percentage of the power supplied to the oven and the cable is wasted in the cable. [2 marks]

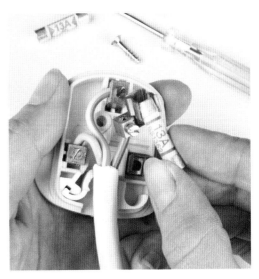

Figure 2 *Changing a fuse*

Worked example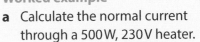

a Calculate the normal current through a 500 W, 230 V heater.

b Determine which fuse, 1 A, 3 A, 5 A, or 13 A, you would use for the appliance.

Solution

a Current $= \dfrac{500\,\text{W}}{230\,\text{V}} = 2.2\,\text{A}$

b You would use a 3 A fuse because it would not melt when the current is 2.2 A, but it would melt if, due to a fault, the current exceeded 3 A. The 5 A and 13 A fuses would only melt if the current exceeded 5 A and 13 A, respectively.

Key points

- The power supplied to a device is the energy transferred to it each second.
- The energy transferred to a device is $E = P \times t$.
- The electrical power supplied to an appliance is equal to $P = I \times V$.
- The correct rating (A) for a fuse:
  $$= \frac{\text{electrical power (watts)}}{\text{potential difference (volts)}}$$

P5.4 Electrical currents and energy transfer

Learning objectives

After this topic, you should know:

- how to calculate the flow of electric charge when you know the current and time
- what energy transfers happen when electrical charge flows through a resistor
- how the energy transferred by a flow of electrical charge is related to potential difference
- about the energy supplied by the battery in a circuit and the energy transferred to the electrical components.

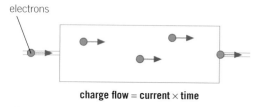

charge flow = current × time

Figure 1 *Charge and current*

Synoptic link

For more information on calculating the energy supplied to an electrical device, look at Topic P1.9.

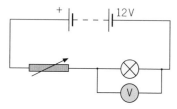

Figure 2 *Energy transfer in a circuit*

Calculating charge

When an electrical appliance is turned on, electrons are forced through it by the potential difference of its power supply unit. The potential difference causes charge to flow through the appliance, carried by the electrons.

You can calculate the charge flow using the equation:

$$\text{charge flow, } Q \quad = \quad \text{current, } I \quad \times \quad \text{time, } t$$
$$\text{(coulombs, C)} \qquad \text{(amperes, A)} \qquad \text{(seconds, s)}$$

Energy and potential difference

When a resistor is connected to a battery, work is done by the battery to make electrons pass through the resistor. Each electron repeatedly collides with the vibrating metal ions of the resistor, transferring energy to the ions. So the ions of the resistor gain kinetic energy and vibrate even more. The resistor becomes hotter. The electrical work done by the battery is equal to the energy transferred to the resistor. In this way, energy is transferred from the chemical energy store in the battery to the resistor's store of thermal energy.

When charge flows through a resistor, energy is transferred to the resistor, so the resistor becomes hotter.

The energy transferred to a resistor E, in a given time t, can be calculated using either of the equations:

energy, E = charge flow, Q × potential difference, V

energy, E = power, P × time, t = potential difference, V × current, I × time, t

Energy transfer in a circuit

The circuit in Figure 2 shows a 12 V battery in series with a torch bulb and a variable resistor. When the voltmeter reads 10 V, the potential difference across the variable resistor is 2 V.

Each coulomb of charge:

- leaves the battery with 12 J of energy (because energy from the battery = charge × battery potential difference)
- transfers 10 J of energy to the torch bulb (because energy transfer to bulb = charge × potential difference across bulb)
- transfers 2 J of energy to the variable resistor.

The energy transferred to the bulb and the resistor increases their thermal energy stores. As a result, the bulb becomes hot and emits light and the variable resistor becomes warm so it heats the surroundings. So energy is transferred to the surroundings by both the bulb and the resistor.

Therefore, the energy from the battery = the energy transferred to the bulb + the energy transferred to the variable resistor.

Worked example

Calculate the energy transferred in a component when the charge passing through it is 30 C and the potential difference is 20 V.

Solution

Using the equation $E = V \times Q$ gives:

energy transferred = 20 V × 30 C = **600 J**

1 **a** Calculate the charge flowing in 50 s when the current is 3 A.
[2 marks]

   b Calculate the energy transferred when the charge flow is 30 C and the potential difference is 4 V.
[2 marks]

   c Calculate the energy transferred in 60 s when a current of 0.5 A passes through a 12 Ω resistor.
[2 marks]

2 **a** Calculate the charge flow for:

    i a current of 4 A for 20 s
[2 marks]

    ii a current of 0.2 A for 60 minutes.
[2 marks]

   b Calculate the energy transfer:

    i for a charge flow of 20 C when the potential difference is 6.0 V
[2 marks]

    ii for a current of 3 A that passes through a resistor for 20 s, when the potential difference is 5 V.
[2 marks]

3 In Figure 2, an ammeter is connected in the circuit in series with the battery. The variable resistor is then adjusted until the ammeter reading is 2.0 A. The voltmeter reading is then 9.0 V.

   a Calculate the charge that passes through the battery in 60 s.
[2 marks]

   b Calculate the energy transferred to or from each coulomb of charge when it passes through each component including the battery.
[3 marks]

   c Show that the energy transferred from the battery in 60 s is equal to the sum of the energy transferred to the lamp and the variable resistor in this time.
[2 marks]

4 In Figure 3 a 4.0 Ω resistor and an 8.0 Ω resistor in series with each other are connected to a 6.0 V battery.

Calculate:

   a the resistance of the two resistors in series
[2 marks]

   b the current through the resistors
[2 marks]

   c the potential difference across each resistor
[2 marks]

   d the energy transferred to each resistor in 60 seconds
[2 marks]

   e the energy supplied by the battery in 60 seconds.
[2 marks]

Study tip

Make sure you know and can use the relationship between charge, current, and time.

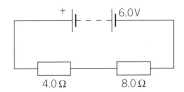

Figure 3 *See summary question 4*

Key points

- The charge flow is $Q = I \times t$.
- When charge flows through a resistor, energy transferred to the resistor makes it hot.
- The energy transferred to a component is $E = V \times Q$.
- When charge flows around a circuit for a given time, the energy supplied by the battery is equal to the energy transferred to all the components in the circuit.

P5.5 Appliances and efficiency

Learning objectives

After this topic, you should know:

- how to calculate the energy supplied to an electrical appliance from its current, its potential difference, and how long it is used for
- how to work out the useful energy output of an electrical appliance
- how to work out the output power of an electrical appliance
- how to compare different appliances that do the same job.

Rearranging the equation

$$E = P\,t$$

To make power the subject, divide both sides by t. this gives:

$$P = \frac{E}{t}$$

Study tip

Remember that 1 kW = 1000 W

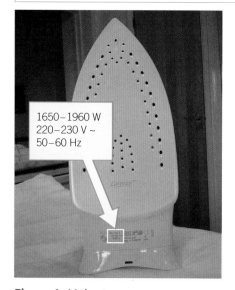

1650–1960 W
220–230 V ~
50–60 Hz

Figure 1 *Mains power*

Synoptic link

You met efficiency and power in Topic P1.9.

When you use an electric heater, how much energy is transferred from the mains? You can work this out if you know its power and how long you use it for.

For any appliance, the energy supplied to it depends on how long it has been switched on and the power supplied to it. For example:

- A 1 kilowatt heater uses the same amount of energy in 1 hour as a 2 kilowatt heater would use in half an hour.
- A 100 W lamp uses the same amount of energy in 30 hours as a 3000 W heater does in 1 hour.

You can use the equation below to work out the energy, in joules, transferred to a mains appliance in a given time:

energy transferred from the mains, E = **power, P** × **time, t**
(joules, J)   (watts, W)   (seconds, s)

To calculate the power supplied to an electrical appliance use the following equation:

power, P = **current, I** × **potential difference, V**
(watts, W)   (amperes, A)   (volts, V)

Electrical appliances and efficiency

The efficiency of any device can be calculated from the power supplied to it (its input power) and the useful energy per second it transfers (its output power) using the equation:

$$\text{efficiency} = \frac{\text{its output power}}{\text{its input power}} (\times 100)$$

Given the input power and efficiency of an appliance, the output power can be calculated by rearranging the above equation to give:

$$\text{output power} = \text{efficiency} \times \text{its input power}$$

Efficiency values can be expressed as a ratio or a percentage. For example, a percentage efficiency of 60% is equivalent to a ratio of 0.60.

The percentage efficiency of an electrical appliance is always less than 100%. The difference between the percentage efficiency of an appliance and 100% tells you the percentage of the energy supplied that is wasted.

Electrical appliances waste energy because the current in both the wires and the components of the appliance has a heating effect due to the resistance of the wires and the components. So they transfer energy by heating to the surroundings. Electrical devices with moving parts such as electric motors also waste energy due to friction between their moving parts. Friction between the moving parts heats them, so they also transfer energy by heating the surroundings.

66

Worked example

A 230 V, 12 A electric motor in a machine has an efficiency of 60%. Calculate:

a the electrical power supplied to it

b the output power of the motor

c the energy per second wasted by the motor.

Solution

a power supplied = current × potential difference = 12 A × 230 V = 2760 W

b efficiency as a ratio = 60% ÷ 100% = 0.60
output power = efficiency × input power = 0.60 × 2760 W = 1660 W

c energy wasted per second = 2760 W − 1660 W = 1100 W

---

1 a Calculate how many joules of energy are used by:
  i a 5 W torch lamp in 50 minutes (3000 seconds) [2 marks]
  ii a 100 W lamp in 24 hours. [2 marks]

 b Calculate how much energy is transferred in each case below.
  i A 3 kilowatt electric kettle is used six times for 5 minutes each time. [2 marks]
  ii A 1000 watt microwave oven is used for 30 minutes. [2 marks]

2 Use the data in Table 1 to answer the following questions about different types of electric lamps.

Table 1

Type	Power in watts	Efficiency
Filament lamp	100	20%
Halogen bulb	100	25%
Low-energy compact fluorescent lamp bulb (CFL)	25	80%
Low-energy light-emitting diode (LED)	2	90%

 a How much energy is wasted in 1 second by:
  i the filament lamp? [1 mark] ii the CFL bulb? [1 mark]

 b A student looks at Table 1 and claims that 50 LEDs would be needed to give the same output as the halogen bulb. Explain why this claim is incorrect and estimate how many LEDs would be needed for this purpose. [3 marks]

3 An electric heater is left on for 4 hours.
 During this time it uses 36 MJ of energy.
 a Calculate the power of the heater. [2 marks]
 b Calculate how long it would take for a 2000 W electric kettle to use 36 MJ of energy. [2 marks]

4 The mains power supply of a computer provides a current of 1.5 A at 230 V.
 a Calculate the power supplied to the computer. [2 marks]
 b In one month, the computer is used for 130 hours. Calculate how many joules of energy are supplied to the computer in this time. [2 marks]

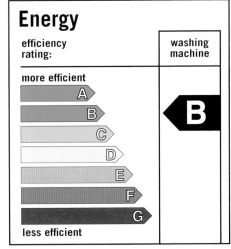

Figure 2 *Efficiency measures. All new appliances such as washing machines and freezers sold in the EU are labelled clearly with an efficiency rating. The rating is from A (very efficient) to G (lowest efficiency). Light bulbs are also labelled in this way on the packaging*

Key points

- A domestic electricity meter measures how much energy is supplied.
- The energy supplied to an appliance is $E = P t$.
- Useful energy used = efficiency × energy supplied.

P5 Electricity in the home

Summary questions

1 a In a mains circuit, name the wire that:
  i is earthed at the local substation [1 mark]
  ii alternates in potential. [1 mark]
  b An oscilloscope is used to display the potential difference of an alternating potential difference supply unit. Write and explain how the trace would change if:
  i the potential difference is increased [2 marks]
  ii the frequency is increased. [2 marks]

2 a Give a reason why a mains appliance with a metal case is unsafe if the case is not earthed. [2 marks]
  b Write the colour of each wire in a mains circuit. [1 mark]

3 a Explain why the wall sockets in a room are connected in parallel with each other [1 mark]
  b A small hairdryer has a double-insulated plastic case and is fitted with a plastic mains cable that contains two wires that are insulated separately.
  i Name the two wires in the cable. [1 mark]
  ii Why are the two wires in the cable insulated separately? [3 marks]
  iii The mains plug of the hairdryer has three terminals. Why does the cable not contain three wires (one for each terminal)? [2 marks]

4 a i Calculate the current in a 230 V, 2.5 kW electric kettle. [2 marks]
  ii Write the fuse, 3 A, 5 A, or 13 A, that would you fit in the kettle plug. [1 mark]
  iii If the kettle is used on average six times a day for 5 minutes each time, calculate the energy in kWh it uses in 28 days. [2 marks]
  b A student uses a 4.0 A, 230 V microwave oven for 10 minutes every day, and a 2500 W electric kettle three times a day for 4 minutes each time. Explain which appliance uses more energy in one day. [3 marks]
  c The electric kettle takes 300 s to heat 1.5 kg of water from 15 °C to 100 °C. Calculate its efficiency. The specific heat capacity of water is 4200 J/kg°C. [4 marks]

5 A 5 Ω resistor is in series with a bulb, a switch and a 12 V battery.
  a Draw the circuit diagram. [1 mark]
  b When the switch is closed for 60 seconds, a direct current of 0.6 A passes through the resistor. Calculate:

  i the energy supplied by the battery [2 marks]
  ii the energy transferred to the resistor [2 marks]
  iii the energy transferred to the bulb. [2 marks]
  c The bulb is replaced by a 25 Ω resistor.
  i Calculate the total resistance of the two resistors. [1 mark]
  ii Calculate the current in the battery. [2 marks]
  iii Calculate the power supplied by the battery and the power delivered to each resistor. [3 marks]

6 A 12 V, 36 W bulb is connected to a 12 V supply.
  a Calculate:
  i the current through the bulb [2 marks]
  ii the charge flow through the bulb in 200 s. [2 marks]
  b i Show that 7200 J of energy is delivered to the bulb in 200 s. [2 marks]
  ii Calculate the energy delivered to the bulb by each coulomb of charge that passes through it. [2 marks]
  c A second 12 V, 36 W bulb is connected to the power supply in parallel with the first bulb.
  i Calculate the current through each bulb and through the battery. [2 marks]
  ii Show that the energy delivered per second to the two bulbs is equal to the energy supplied per second by the battery. [3 marks]

7 An electrician has the job of connecting a 6.6 kW electric oven to the 230 V mains supply in a house.
  a Calculate the current needed to supply 6.6 kW of electrical power at 230 V. [2 marks]
  b The table below shows the maximum current that can pass safely through five different mains cables. For each cable the cross-sectional area of each conductor is given in square millimetres (mm²).

Table 1

	Cross-sectional area of conductor in millimetres squared	Maximum safe current in amperes
A	1.0	14
B	1.5	18
C	2.5	28
D	4.0	36
E	6.0	46

  i To connect the oven to the mains supply, determine which cable the electrician should choose. Give a reason for your answer. [3 marks]
  ii Explain what would happen if she chose a cable with thinner conductors. [2 marks]

Practice questions

01.1 Complete the sentences using the words in the box. Each word can be used once, more than once, or not at all.

| electrons | protons | current | the same |
| a changing | volts | | |

When an electric _____ flows through a wire it is a flow of _____. [2 marks]

In a d.c. circuit the flow is always in _____ direction. [1 mark]

In an a.c. circuit the flow is always in _____ direction. [1 mark]

01.2 An oscilloscope is connected to a power supply.

Figure 1

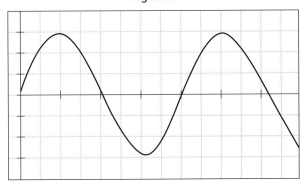

Determine the peak voltage and frequency of the supply. The Y-gain control is set at 6.5 volts/division and the time base is set at 10 milliseconds/division. [2 marks]

02.1 Choose two boxes below that describe the National Grid. [2 marks]

A system of pylons and cables across the country. ☐

A system linking power stations to consumers. ☐

A system of crossed structures on pylons ☐

A system of cables and transformers ☐

A system linking power stations to houses ☐

02.2 Describe the difference in use between a step-up and a step-down transformer. [2 marks]

02.3 Explain why it is advantageous to have a step up transformer in the system. [3 marks]

03 A student investigates the power rating of a resistor. The apparatus used is shown in **Figure 2**.

Figure 2

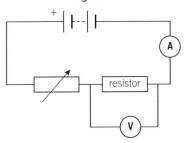

03.1 Give **one** reason why a variable resistor is used in the circuit. [1 mark]

03.2 Give the resolution of the voltmeter. [1 mark]

Figure 3

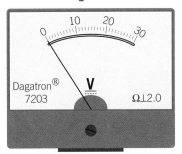

03.3 Calculate the power of the fixed resistor. The potential difference across the resistor is 12 V and the current through the resistor is 50 mA. [3 marks]

03.4 Calculate the resistance of the fixed resistor. Use the information in **03.3**. [2 marks]

04 A rating plate for an electric toaster states: 230 volts, 50 Hz, 800 watts.

04.1 Give the frequency of the mains supply. [1 mark]

04.2 Give the power rating of the toaster. [1 mark]

04.3 Calculate the current in amps flowing through the toaster. [2 marks]

04.4 Calculate the energy supplied to the toaster in 2.5 minutes. [2 marks]

04.5 Calculate the amount of charge flowing through the toaster if it takes 2.5 minutes to toast two slices of bread. Give your answer to 3 significant figures. [3 marks]

05 A gardener uses a plastic electric lawn mower to cut the grass.

05.1 Describe what is meant in the handbook by "the lawn mower is double insulated". [2 marks]

05.2 Explain how the gardener may get an electric shock if the cable gets cut whilst using the lawnmower. [3 marks]

Learning objectives

After this topic, you should know:

- how density is defined and its units of measurement
- how to measure the density of a solid object or a liquid
- how to use the density equation to calculate the mass or the volume of an object or a sample
- how to tell from its density if an object will float in water.

Figure 1 *Materials of different densities*

Rearranging equations

Rearranging the density equation $\rho = \dfrac{m}{V}$ gives:

$$m = \rho V \qquad \text{or} \qquad V = \dfrac{m}{\rho}$$

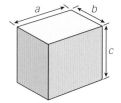

volume of cuboid = $a \times b \times c$

Figure 2 *The volume of a cuboid*

Density comparisons

Any builder knows that a concrete post is much heavier than a wooden post of the same size. This is because the **density** of concrete is much greater than the density of wood. A volume of one cubic metre of wood has a mass of about 800 kg. But a cubic metre of concrete has a mass of about 2400 kg. So the density of concrete is about three times the density of wood.

The density of a substance is defined as its mass per unit volume.

You can use the equation below to calculate the density ρ of a substance if you know the mass m and the volume V of a sample of it.

$$\underset{\text{(kilogram per cubic metre, kg/m}^3)}{\text{density, } \rho} = \underset{\text{volume, } V \text{ (metres}^3, \text{ m}^3)}{\dfrac{\text{mass, } m \text{ (kilograms, kg)}}{}}$$

Converting units and using standard form

1 kg = 1000 g = 10^3 g

1 m = 100 cm = 10^2 cm

1 m³ = 1 000 000 cm³ = 10^6 cm³

So 1000 kg/m³ = 1 000 000 g/1 000 000 cm³ = 1 g/cm³

Standard form is useful when you are working with very large numbers, particularly when you need to convert values to SI units (e.g., converting MJ to J) for a calculation. In standard form, a number is written as $A \times 10^n$, where n is the number of places you have had to move the decimal point to the left (or right for a negative power of ten) to get the decimal number A, which is greater than 1 and less than 10.

Worked example

A wooden post has a volume of 0.025 m³ and a mass of 20 kg. Calculate its density in kg/m³.

Solution

$$\text{density} = \dfrac{\text{mass}}{\text{volume}} = \dfrac{20 \text{ kg}}{0.025 \text{ m}^3} = 800 \text{ kg/m}^3$$

Measuring the density of a solid object

To measure the mass of the object, use an electronic balance. Make sure the balance reads zero before you place the object on it.

To find the volume of a regular solid, such as a cube or a cuboid, measure its dimensions using a millimetre ruler, vernier callipers, or a micrometer – whichever is the most appropriate. Use the measurements and the equation shown in Figure 2 to calculate its volume.

For a small irregular solid, lower it on a thread into a measuring cylinder partly filled with water. You can work out the volume of the object by the rise in the water level.

Measuring the density of a liquid

Use a measuring cylinder to measure the volume of a particular amount of the liquid.

Measure the mass of an empty beaker using a balance. Remove the beaker from the balance and pour the liquid from the measuring cylinder into the beaker. Use the balance again to measure the total mass of the beaker and the liquid. You can calculate the mass of the liquid by subtracting the mass of the empty beaker from the total mass of the beaker and the liquid.

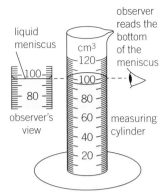

Figure 3 *Using a measuring cylinder*

Worked example

A measuring cylinder contained a volume of 120 cm³ of a particular liquid. The liquid was then poured into an empty beaker of mass 51 g. The total mass of the beaker and the liquid was then found to be 145 g.

a Calculate the mass of the liquid in grams.

b Calculate the density of the liquid in kg/m³.

Solution

a Mass of liquid = 145 − 51 = **94 g.**

b $\text{density} = \dfrac{\text{mass}}{\text{volume}} = \dfrac{94\,g}{120\,cm^3} = \dfrac{0.094\,kg}{0.000120\,m^3} = \mathbf{780\,kg/m^3}$

---

1 A rectangular concrete slab is 0.80 m long, 0.60 m wide, and 0.05 m thick.
   a Calculate its volume in m³. [1 mark]
   b The mass of the concrete slab is 60 kg. Calculate its density in kg/m³. [2 marks]

2 A measuring cylinder contains 80 cm³ of a particular liquid. The liquid is poured into an empty beaker of mass 48 g. The total mass of the beaker and the liquid was found to be 136 g.
   a Calculate the mass of the liquid in grams. [2 marks]
   b Calculate the density of the liquid in g/cm³. [2 marks]

3 A rectangular block of gold is 0.10 m in length, 0.08 m in width, and 0.05 m in thickness.
   a i Calculate the volume of the block. [1 mark]
      ii The mass of the block is 0.76 kg. Calculate the density of gold. [2 marks]
   b A thin gold sheet has a length of 0.15 m and a width of 0.12 m. The mass of the sheet is 0.0015 kg. Use these measurements and the result of your density calculation in part **a ii** to calculate the thickness of the sheet. [3 marks]

4 Describe how you would measure the density of a metal bolt. You may assume the bolt will fit into a measuring cylinder of capacity 100 cm³. [4 marks]

Density tests

For each of the tests, measure the mass and the volume of the object as explained. Then use the equation

$\text{density} = \dfrac{\text{mass}}{\text{volume}}$ to calculate the density of the object.

Safety: Take care not to spill any liquids and, if you do, let your teacher know.

Study tip

The instrument you choose to use to take a measurement is important – you should consider the resolution and range.

Instrument	resolution	range
metre rule mm scale	±0.5 mm	1 m
vernier callipers	±0.05 mm	about 100 mm
micrometer	±0.005 mm	about 30 mm

Key points

- $\text{density} = \dfrac{\text{mass}}{\text{volume}}$ (in kg/m³)
- To measure the density of a solid object or a liquid, measure its mass and its volume, then use the density equation $\rho = \dfrac{m}{V}$.
- Rearranging the density equation gives $m = \rho V$ or $V = \dfrac{m}{\rho}$
- Objects that have a lower density than water (i.e., < 1000 kg/m³) float in water.

P6.2 States of matter

Learning objectives

After this topic, you should know:

- the different properties of solids, liquids, and gases
- the arrangement of particles in a solid, a liquid, and a gas
- why gases are less dense than solids and liquids
- why the mass of a substance that changes state stays the same.

Everything around you is made up of matter and exists in one of three states – solid, liquid, or gas. The table below summarises the main differences between the three states of matter.

Table 1 *Differences between the three states of matter*

State	Flow	Shape	Volume	Density
solid	no	fixed	fixed	much higher than a gas
liquid	yes	fits container shape	fixed	much higher than a gas
gas	yes	fills container	can be changed	lower than a solid or a liquid

Change of state

A substance can change from one state to another, as shown in Figure 2. Changes of state are examples of **physical changes** because no new substances are produced. If a physical change is reversed, the substance recovers its original properties. You can change the state of a substance by heating or cooling the substance.

For example:

- when water in a kettle boils, the water turns into steam. Steam (also called water vapour) is water in its gaseous state
- when solid carbon dioxide (also called dry ice) warms up, the solid turns into gas directly
- when steam touches a cold surface, the steam condenses and turns into water.

Conservation of mass

When a substance changes state, the number of particles in the substance stays unchanged. So the mass of the substance after the change of state is the same as the mass of the substance before the change of state.

In other words, the mass of the substance is conserved when it changes its state.

For example:

- when a given mass of ice melts, the water it turns into has the same mass. So the mass of the substance stays unchanged.
- when water is boiled in a kettle and some of the water turns into steam, the mass of the steam produced is the same as the mass of water boiled away. So the mass of the substance is unchanged even though some of it (i.e., the steam) is no longer in the kettle.

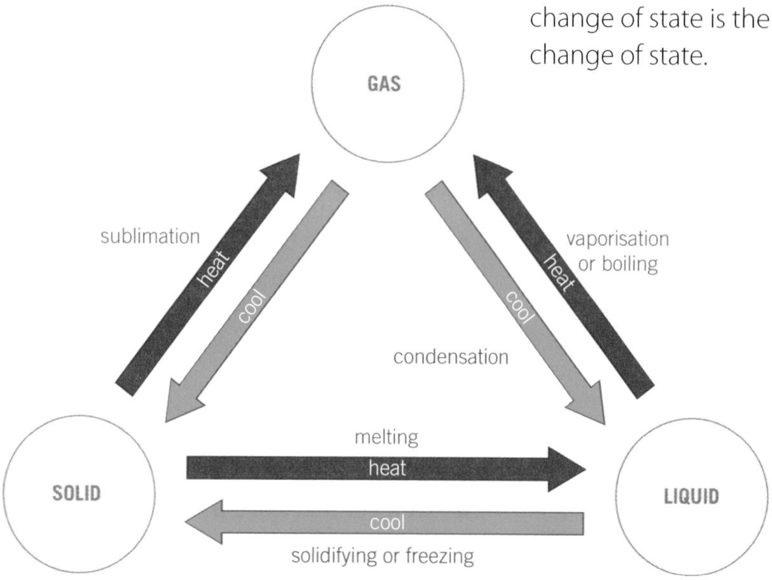

Figure 1 *Spot the three states of matter*

Figure 2 *Change of state*

The kinetic theory of matter

Solids, liquids, and gases are made of particles. Figure 4 shows the arrangement of the particles of a substance in its solid, liquid, and gas states. When the temperature of the substance is increased, the particles move faster.

- The particles of a substance in its solid state are held next to each other in fixed positions. They vibrate about their fixed positions, so the solid keeps its own shape.

- The particles of a substance in its liquid state are in contact with each other. They move about at random. So a liquid doesn't have its own shape, and it can flow.

- The particles of a substance in its gas state move about at random much faster than they do in a liquid. They are, on average, much further apart from each other than the particles of a liquid. So the density of a gas is much less than that of a solid or a liquid.

- The particles of a substance in its solid, liquid, and gas states have different amounts of energy. For a given amount of a substance, its particles have more energy in the gas state than they have in the liquid state, and they have more energy in the liquid state than they have in the solid state.

Figure 4 *The arrangement of particles of a substance in solid state, liquid state, and gas state*

1. Name the change of state that occurs when:
   - **a i** wet clothing on a washing line dries out [1 mark]
   - **ii** hailstones form [1 mark]
   - **iii** snowflakes turn to liquid water. [1 mark]
   - **b** When an ice cube in an empty beaker melts, the volume of water in the beaker just after the ice has melted is less than the volume of the ice cube. Explain what this tells you about the density of ice compared with the density of the water just after the ice has melted. [2 marks]

2. Give the scientific word for each of the following changes:
   - **a** the windows in a bus full of people mist up [1 mark]
   - **b** water vapour is produced from the surface of the water in a pan when the water is heated before it boils [1 mark]
   - **c** ice cubes taken from a freezer thaw out [1 mark]
   - **d** water put into a freezer gradually turns to ice. [1 mark]

3. Describe the changes that take place in the movement and arrangement of the particles in:
   - **a** an ice cube when the ice melts [2 marks]
   - **b** water vapour when it condenses on a cold surface. [3 marks]

4. Explain, using the kinetic theory of matter, why liquids and solids are much denser than gases. [4 marks]

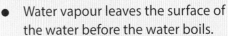
Changing state

Heat some water in a beaker using a Bunsen burner (Figure 3).

- Water vapour leaves the surface of the water before the water boils.

- When the water boils, bubbles of vapour form inside the water and rise to the surface to release steam.

Switch the Bunsen burner off and hold a cold beaker or cold metal object above the boiling water. Observe the condensation of steam from the boiling water on the cold object.

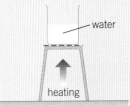

Figure 3 *Heating water to show a change of state*

Safety: Take care with boiling water, and wear eye protection.

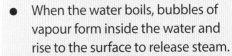

P6.3 Changes of state

Learning objectives

After this topic, you should know:

- what is meant by the melting point and the boiling point of a substance
- what is needed to melt a solid or to boil a liquid
- how to explain the difference between boiling and evaporation
- how to use a temperature–time graph to find the melting point or the boiling point of a substance.

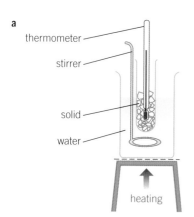

a

thermometer

stirrer

solid

water

heating

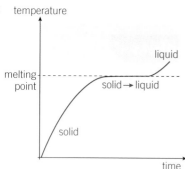

b temperature

liquid

melting point

solid → liquid

solid

time

Figure 1a *Measuring the melting point of a substance* **b** *A temperature–time graph*

Study tip

Don't forget that during the time a pure substance is changing its state, its temperature does *not* change.

Melting points and boiling points

When pure ice is heated and melts, its temperature stays at 0 °C until all the ice has melted. When water is heated and boils at atmospheric pressure, its temperature stays at 100 °C.

For any pure substance undergoing a change of state, its temperature stays the same while the change of state is taking place. The temperature at which a solid changes to a liquid is called the **melting point**, and the temperature at which a liquid turns to a gas is called the **boiling point** of the substance. The temperature at which a liquid changes to a solid is called its **freezing point**. It is the same temperature as the melting point of the solid.

The melting point of a solid and the boiling point of a liquid are affected by impurities in the substance. For example, the melting point of water is lowered if you add salt to the water. This is why salt is added to the grit that's used for gritting roads in freezing weather – it means roads don't get icy until they are colder.

Table 1

Change of state	Initial and final state	Temperature
Melting	solid to liquid	melting point
Freezing (also called solidification)	liquid to solid	
Boiling	liquid to gas	boiling point
Condensation	gas to liquid	

Measuring the melting point of a substance

Place a substance in its solid state in a suitable test tube in a beaker of water (Figure 1a). Heat the water, and measure the temperature of the substance when it melts. If its temperature is measured every minute, you can plot the measurements on a graph (Figure 1b). The melting point is the temperature of the flat section of the graph because this is when the temperature stays the same during the time in which the substance is melting.

You can use the same arrangement without the beaker of water to find the boiling point of a liquid.

Safety: Wear eye protection.

Energy and change of state

Suppose a beaker of ice below 0 °C is heated steadily so that the ice melts and then the water boils. Figure 2 shows how the temperature changes with time.

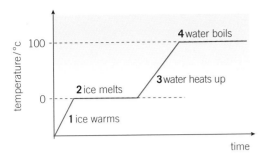

Figure 2 *Melting and boiling water*

The temperature of the water:

1   increases until it reaches 0 °C when the ice starts to melt at 0 °C, then

2   stays constant at 0 °C until all the ice has melted, then

3   increases from 0 °C to 100 °C until the water in the beaker starts to boil at 100 °C, then

4   stays constant at 100 °C as the water turns to steam (until all the water has boiled away, if the water continues to be heated).

The energy transferred to a substance when it changes its state is called **latent heat**. The energy transferred to the substance to melt or boil it is 'hidden' by the substance because its temperature does not change at the substance's melting point or at its boiling point.

Most pure substances produce a temperature–time graph with similar features to Figure 2. Note that:

● fusion is sometimes used to describe melting because different solids can be joined, or 'fused', together when they melt

● evaporation from a liquid happens at its surface when the liquid is below its boiling point. At its boiling point, a liquid boils because bubbles of vapour form inside the liquid and rise to the surface to release the gas.

1   Write three differences between evaporation and boiling.   [3 marks]

2   A pure solid substance X was heated in a tube and its temperature was measured every 30 seconds.
     The measurements are given in the table below.

Time in s	0	30	60	90	120	150	180	210	240	270	300
Temperature in °C	20	35	49	61	71	79	79	79	79	86	92

   a i Use the measurements in the table to plot a graph of temperature (y-axis) against time (x-axis).   [3 marks]
   ii Use your graph to find the melting point of X.   [1 mark]
   b Describe the physical state of the substance as it was heated from 60 °C to 90 °C.   [3 marks]

3   Salt water has a lower freezing point than pure water. In icy conditions in winter, gritting lorries are used to scatter a mixture of salt and grit on roads. Explain the purpose of each of the two components of the mixture.   [3 marks]

4   A substance has a melting point of 75 °C. Describe how the arrangement and motion of the particles changes as the substance cools from 80 °C to 70 °C.   [4 marks]

Study tip

Evaporation happens at any temperature – boiling happens only at the substance's boiling point.

Key points

● For a pure substance:
  ▪ its melting point is the temperature at which it melts (which is the same temperature at which it solidifies)
  ▪ its boiling point is the temperature at which it boils (which is the same temperature at which it condenses).
● Energy is needed to melt a solid or to boil a liquid.
● Boiling occurs throughout a liquid at its boiling point. Evaporation occurs from the surface of a liquid when its temperature is below its boiling point.
● The flat section of a temperature–time graph gives the melting point or the boiling point of a substance.

P6.4 Internal energy

Learning objectives

After this topic, you should know:

- how increasing the temperature of a substance affects its internal energy
- how to explain the different properties of a solid, a liquid, and a gas
- how the energy of the particles of a substance changes when it is heated
- how to explain in terms of particles why a gas exerts pressure.

Synoptic link

For more about energy transfers, look back at Topic P1.1.

Synoptic link

For more about specific heat capacity, look back at Topic P2.2.

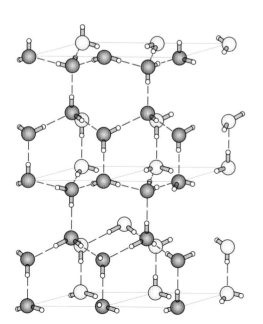

Figure 1 *Molecular model of ice*

When you switch a kettle on, the temperature of the water in the kettle increases until the water boils. The molecules in the water gain energy and move about faster as the temperature of the water increases. When the water boils, it means that the molecules have gained enough energy to move away from each other so that the water turns into vapour (steam).

The energy stored by the particles of a substance is called the substance's **internal energy**. This is the energy of the particles that is caused by their individual motion and positions. The internal energy of the particles is the sum of:

- the kinetic energy they have due to their individual motions relative to each other, and
- the potential energy they have due to their individual positions relative to each other.

So the internal energy of a substance is the total energy in the kinetic and potential energy stores of all the particles in the substance that is caused by their individual motions and positions.

Internal energy does **not** include gravitational potential energy or the kinetic energy that is caused by the motion of the whole substance.

Heating a substance changes the internal energy of the substance by increasing the energy of its particles. Because of this, the temperature of the substance increases or its physical state changes (i.e., it melts or boils).

- When the temperature of a substance increases (or decreases), the total kinetic energy of its particles increases (or decreases). For a given mass m of a substance of specific heat capacity c, the energy E needed to change its temperature by $\Delta\theta$ without a change of state is given by the specific heat equation $\Delta E = m\,c\,\Delta\theta$.
- When the physical state of a substance changes, the total potential energy of its particles changes. As you learnt in Topic P6.3, the term latent heat is used to describe the energy transferred to or from a substance when it changes state.

You'll learn more about latent heat in the next topic.

Comparing the particles in solids, liquids, and gases

In a solid, particles (i.e., atoms and molecules) are arranged in a three-dimensional structure.

- There are strong forces of attraction between these particles. These forces bond the particles in fixed positions.
- Each particle vibrates about an average position that is fixed.
- When a solid is heated, the particles' energy stores increase and they vibrate more. If the solid is heated up enough, the solid melts (or sublimates) because its particles have gained enough energy to break away from the structure.

In a liquid, there are weaker forces of attraction between the particles than in a solid. These weak forces of attraction are not strong enough to hold the particles together in a rigid structure.

Figure 2 *Molecules in water*

- The forces of attraction are strong enough to stop the particles moving away from each other completely at the surface.

- When a liquid is heated, some of the particles gain enough energy to break away from the other particles. The molecules that escape from the liquid are in a gas state above the liquid.

In a gas, the forces of attraction between the particles are so weak, they are insignificant.

- The particles move about at high speed in random directions, colliding with each other and with the internal surface of their container. The **pressure** of a gas on a solid surface such as a container is caused by the force of impacts of the gas particles with the surface.

- When a gas is heated, its particles gain kinetic energy and on average move faster. This causes the pressure of the gas to increase because the particles collide with the container surface more often and with more force.

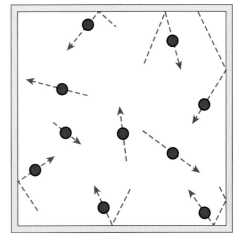

Figure 3 *Gas molecules in a box*

1 Explain the following statements in terms of particles.
   a A gas exerts a pressure on any surface it is in contact with. [3 marks]
   b Heating a solid makes it melt. [3 marks]

2 Table 1 lists the properties of the molecules in four different substances. Write, with a reason, whether each substance is a solid, a liquid, or a gas, or doesn't exist.

   Table 1

	Distance between the molecules	Particle arrangement	Movement of the molecules	
a	close together	not fixed	move about	[1 mark]
b	far apart	not fixed	move about	[1 mark]
c	close together	fixed	vibrate	[1 mark]
d	far apart	fixed	vibrate	[1 mark]

3 Explain why the internal energy of a solid increases when it is heated at its melting point. [2 marks]

4 An ice cube at 0 °C is placed in a beaker of water to cool the water down. Describe the energy changes of the particles of the ice and the water that takes place. [4 marks]

Key points

- Increasing the temperature of a substance increases its internal energy.
- The strength of the forces of attraction between the particles of a substance explains why it is a solid, a liquid, or a gas.
- When a substance is heated:
  - if its temperature rises, the kinetic energy of its particles increases
  - if it melts or it boils, the potential energy of its particles increases.
- The pressure of a gas on a surface is caused by the particles of the gas repeatedly hitting the surface.

P6.5 Specific latent heat

Learning objectives

Study tip

Latent heat is the energy transferred when a substance changes its state.

Specific latent heat is the energy transferred per kilogram when a substance changes its state.

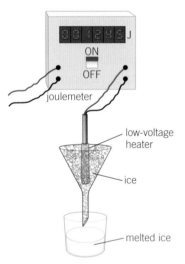

Figure 1 *Measuring the specific latent heat of fusion of ice*

Synoptic links

Instead of using a joulemeter, the energy supplied to the heater can be measured using the circuit and information in Topic P2.2.

Latent heat of fusion

When a solid substance is heated at its melting point, the substance melts and turns into liquid. Its temperature stays constant until all of the substance has melted. The energy supplied is called latent heat of fusion. It is the energy needed by the particles to break free from each other.

If the substance in its liquid state is cooled, it will solidify at the same temperature as its melting point. When this happens, the particles bond together into a rigid structure. Latent heat is transferred to the surroundings as the substance solidifies and the particles form stronger bonds.

The **specific latent heat of fusion** L_F of a substance is the energy needed to change the state of 1 kg of the substance from solid to liquid, at its melting point (i.e. without changing its temperature).

The unit of specific latent heat of fusion is the joule per kilogram (J/kg).

If energy E is transferred to a solid at its melting point, and mass m of the substance melts without change in temperature:

$$\textbf{specific latent heat of fusion, } L_F \text{ (J/kg)} = \frac{\textbf{energy, } E \text{ (joules, J)}}{\textbf{mass, } m \text{ (kilograms, kg)}}$$

You can rearrange this equation to $E = m L_F$

Specific latent heat of fusion of ice

In this experiment, a low-voltage heater is used to melt crushed ice in a funnel. The melted ice is collected using a beaker under the funnel (Figure 1). A joulemeter is used to measure the energy supplied to the heater.

1. With the heater off, water from the funnel is collected in the beaker for a measured time (e.g., 10 minutes). The mass of the beaker and water m_1 is then measured. The beaker is then emptied for the next stage.

2. With the heater on, the procedure is repeated for exactly the same time. The joulemeter readings before and after the heater is switched on are recorded. After the heater is switched off, the mass of the beaker and the water m_2 is measured once more.

To calculate the specific latent heat of fusion of ice, note that:

- the mass of ice melted because of the heater is $m = m_2 - m_1$
- the energy supplied E to the heater = the difference between the joulemeter readings
- the specific latent heat of fusion of ice is $L_F = \dfrac{E}{m} = \dfrac{E}{m_2 - m_1}$.

Safety: Take care with a hot immersion heater, and wear eye protection.

Latent heat of vaporisation

When a liquid substance is heated, at its boiling point, the substance boils and turns into vapour. The energy supplied is called latent heat of vaporisation. It is the energy needed by the particles to break away from their neighbouring particles in the liquid.

If the substance in its gas state is cooled, it will condense at the same temperature as its boiling point. Latent heat is transferred to the surroundings as the substance condenses into a liquid and its particles form new bonds.

The **specific latent heat of vaporisation** L_v of a substance is the energy needed to change the state of 1 kg of the substance from liquid to vapour, at its boiling point (i.e., without changing its temperature).

The unit of specific latent heat of vaporisation is the joule per kilogram (J/kg).

If energy E is transferred to a liquid at its boiling point, and mass m of the substance boils away without change in temperature:

$$\text{specific latent heat of vaporisation, } L_v = \frac{\text{energy, } E \text{ (joules, J)}}{\text{mass, } m \text{ (kilograms, kg)}}$$
(joules per kilogram, J/kg)

1 In the experiment shown in Figure 1, 0.024 kg of water was collected in the beaker in 300 s with the heater turned off. The beaker was then emptied and placed under the funnel again. With the heater on for exactly 300 s, the joulemeter reading increased from zero to 15 000 J, and 0.068 kg of water was collected in the beaker.
   a Calculate the mass of ice melted because of the heater being on. [1 mark]
   b Use the data to calculate the specific latent heat of fusion of water. [2 marks]

2 In the experiment shown in Figure 2, the balance reading decreased from 0.152 kg to 0.144 kg in the time taken to supply 18 400 J of energy to the boiling water. Use the data to calculate the specific latent heat of vaporisation of water. [3 marks]

3 An ice cube of mass 0.008 kg at 0 °C was placed in water at 15 °C in an insulated plastic beaker. The mass of water in the beaker was 0.120 kg. After the ice cube had melted, the water was stirred, and its temperature was found to have fallen to 9 °C. The specific heat capacity of water is 4200 J/kg °C.
   a Calculate the energy transferred from the water. [2 marks]
   b Show that when the melted ice warmed from 0 °C to 9 °C, it gained 300 J of energy. [2 marks]
   c Use this data to calculate the specific latent heat of fusion of water. [3 marks]

4 Estimate how long a 3000 W electric kettle would take to boil away 100 g of water. The specific latent heat of vapourisation of water is 2.25 MJkg. [3 marks]

Specific latent heat of vaporisation of water

Use a low-voltage heater (Figure 2) to bring water in an insulated beaker to the boil. The joulemeter reading and the top pan balance reading are then measured and then remeasured after a certain time (e.g., 5 minutes).

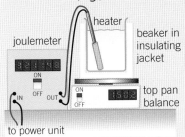

Figure 2 *Measuring the specific latent heat of vaporisation of water*

In this time:
- the energy supplied E = the difference between the joulemeter readings.
- the mass of water boiled away, m = the difference between the readings of the top pan balance.
- the specific latent heat of vaporisation of water is $L_v = \frac{E}{m}$.

Safety: This experiment will be demonstrated by your teacher. You should wear eye protection and stand behind a safety screen. Your teacher will need to take care with the hot immersion heater.

Key points

- Latent heat is the energy needed for a substance to change its state without changing its temperature.
- Specific latent heat of fusion (or of vaporisation) is the energy needed to melt (or to boil) 1 kg of a substance without changing its temperature.
- In latent heat calculations, use the equation $E = m\,L$.
- The specific latent heat of ice (or of water) can be measured using a low-voltage heater to melt the ice (or to boil the water).

P6.6 Gas pressure and temperature

Learning objectives

After this topic, you should know:

- how a gas exerts pressure on a surface
- how the pressure of a gas in a sealed container is affected by changing the temperature of the gas
- how to see evidence of gas molecules moving around at random.

Go further!

Pressure is force per unit area acting on a surface perpendicular to the surface. The unit of pressure is the pascal (Pa). One pascal is the pressure when a force of 1 Newton acts on a surface area of 1 square metre.

In the kitchen

Never heat food in a sealed can. The can will probably explode because the pressure of gas inside it increases as the temperature increases. This is because the molecules of gas in the can collide repeatedly with each other and with the surface inside their container, rebounding after each collision. Each impact with the surface exerts a tiny force on the surface. Millions of millions of these impacts happen every second, and together the total force causes a steady pressure on the surface inside the container. The pressure of a gas on a surface is the total force exerted on a unit area of the surface.

Increasing the temperature of any sealed gas container increases the pressure of the gas inside it. This is because:

- the energy transferred to the gas when it's heated increases the kinetic energy of its molecules. So the average kinetic energy of the gas molecules increases when the temperature of the gas is increased.

- the average speed of the molecules increases when the kinetic energy increases, and the molecules on average hit the container surfaces with more force and more often. So the pressure of the gas increases.

Gas pressure and temperature

Figure 1 shows dry air in a sealed flask connected to a pressure gauge. The flask is in a big beaker of water, which is heated to raise the temperature of the gas. The water is heated in stages to raise the temperature in stages. At each stage, the water is stirred to make sure that its temperature is the same throughout. The temperature of the water is measured using the thermometer. The pressure is read off the pressure gauge.

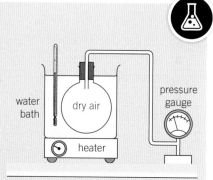

Figure 1 *Measuring gas pressure at different temperatures*

- If the measurements are plotted on a graph of pressure against temperature in °C, the results give you a straight-line graph as shown in Figure 2. This shows that the increase of pressure is the same for equal increases of temperature.

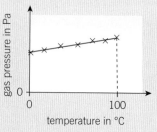

Figure 2 *Pressure–temperature graph for a gas*

Safety: This experiment should be carried out by your teacher and behind a safety screen. You should wear safety goggles.

Observing random motion

Individual molecules are too small for you to see directly. But you can see the effects of them by observing the motion of smoke particles in air. Figure 3 shows how you can do this using a smoke cell and a microscope. The smoke particles move about haphazardly and follow unpredictable paths.

1 A small glass cell is filled with smoke

2 Light is shone through the cell

3 The smoke is viewed through a microscope

4 You see the smoke particles constantly moving and changing direction. The path taken by one smoke particle will look something like this

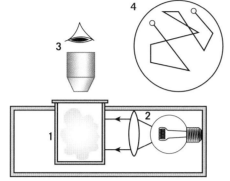

Figure 3 *A smoke cell*

Figure 4 shows how the random motion of smoke particles in air happens. Air molecules repeatedly collide at random with each smoke particle. The air molecules must be moving very fast to make this happen, because they are much too small to see, and the smoke particles are much, much bigger than the air molecules. What you see is the random motion of the smoke particles caused by the random impacts that the gas (air) molecules make on each smoke particle.

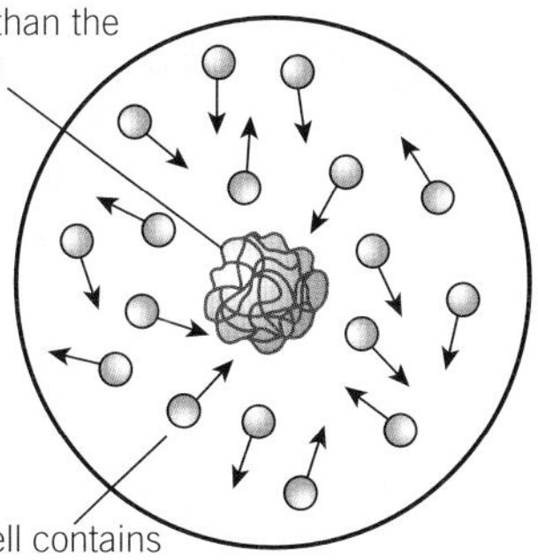

The smoke particle is much larger than the air molecules

The glass cell contains air molecules that are in constant erratic motion. As they collide with the smoke particle they give it a push. The direction of the push changes at random

Figure 4 *The random motion of smoke particles*

1 When a gas is heated in a sealed container, write how, if at all, each of the following properties of the gas changes:
   a The pressure of the gas [1 mark]
   b The average separation of the molecules [1 mark]
   c The number of impacts the molecules make on the surface of the container each second. [1 mark]

2 Explain why smoke particles in air move about faster if the temperature of the air is increased. [3 marks]

3 A gas cylinder is fitted with a valve that opens and lets gas out if the gas becomes too hot. Explain how the gas pressure changes if the gas becomes too hot and the valve opens. 🔩 [3 marks]

4 Look back at the practical on the previous page.
   a Explain why the water must be stirred before its temperature is measured. [2 marks]
   b Explain why the pressure gauge does not read zero before the water is heated. [2 marks]

Key points

- The pressure of a gas is caused by the random impacts of gas molecules on surfaces that are in contact with the gas.
- If the temperature of a gas in a sealed container is increased, the pressure of the gas increases because:
  - the molecules move faster so they hit the surfaces with more force
  - the number of impacts per second of gas molecules on the surfaces of a sealed container increases, so the total force of the impacts increases.
- The unpredictable motion of smoke particles is evidence of the random motion of gas molecules.

P6 Molecules and matter

Summary questions

1 In a paint factory, empty steel tins of mass 0.320 kg and volume 0.001 m³ are filled with paint of density 2500 kg/m³.

   a Calculate the mass of paint in each filled paint tin. [2 marks]

   b Calculate the total weight in newtons of each filled paint tin. [2 marks]

2 This question is about A4 paper for use in a photocopier.

   a Use a millimetre ruler to measure the length and the width of a sheet of this paper. [1 mark]

   b The paper has a mass per unit area of 80 g/m². Calculate the mass of a single sheet of the paper. [2 marks]

   c A packet of 500 sheets of this paper has a thickness of 50 mm. Calculate the thickness of a single sheet of the paper. [1 mark]

   d Use your answers to **b** and **c** to calculate the density of the paper in

     i g/cm³ [3 marks]

     ii kg/m³ [1 mark]

3 A test tube containing a solid substance is heated in a beaker of water. The temperature–time graph shows how the temperature of the substance changed with time as it was heated.

Figure 1

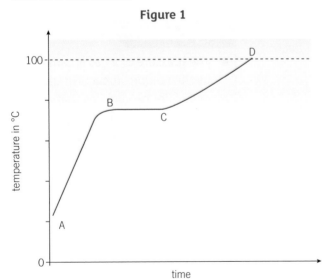

   a Explain why the temperature of the substance:

     i increased from **A** to **B** [1 mark]

     ii stayed the same from **B** to **C** [1 mark]

     iii increased from **C** to **D**. [1 mark]

   b Use the graph to estimate the melting point of the solid. [1 mark]

   c Describe how the arrangement and motion of the particles changes as the temperature increases from **A** to **D**. [3 marks]

4 **a** A plastic beaker containing 0.10 kg of water at 18 °C was placed in a refrigerator for 450 seconds. After this time, the temperature of the water was found to be 3 °C. The specific heat capacity of water is 4200 J/kg °C.

     i Calculate the energy transferred from the water. [2 marks]

     ii Calculate the rate of transfer of energy from the water. [2 marks]

   b **i** Calculate how much more energy would need to be removed from the water to cool it from 3 °C and to freeze it. The specific latent heat of fusion of water is 340 kJ/kg. [4 marks]

     ii Estimate how long it would take to cool it from 3 °C and freeze it. [2 marks]

5 A 3.0 kW electric kettle is fitted with a safety cut-out designed to switch it off as soon as the water boils. Unfortunately, the cut-out does not operate correctly and allows the water to boil for 30 seconds longer than it is supposed to.

   a Calculate how much energy is supplied to the kettle in this time. [2 marks]

   b The specific latent heat of vaporisation of water is 2.3 MJ/kg. Estimate the mass of water boiled away in this time. [2 marks]

6 In a chemistry experiment, 25 cm³ of a gas is collected in a syringe at a pressure of 120 kPa.

   a The density of the gas was 0.0018 g/cm³ at the temperature at which it was collected. Calculate the mass of gas collected. [2 marks]

   b If the gas in **a** was then cooled without changing its volume, write and explain how its pressure would change. 🚫 [4 marks]

7 **a** Explain in terms of particles why gases have a much lower density than solids and liquids. 🚫 [3 marks]

   b Describe how the arrangement and motion of the particles of a substance change when the substance changes its state from liquid to solid. [4 marks]

Practice questions

01 A large statue stands on a square base at the side of a garden pond. Both the base and the statue are made of granite.

01.1 Describe how the volume of the base can be calculated without moving it. [1 mark]

01.2 The statue falls into the pond. A gardener states that the volume of the statue can be estimated by measuring the new water level. Explain if this statement is correct. [2 marks]

01.3 Calculate whether a mechanical hoist with a maximum lift force of 750 N can be used to lift the statue out of the pond. The volume of the statue is 0.027 m³ and the density of granite is 2800 kg/m³. Gravitational field strength = 9.8 N/kg.

[3 marks]

02 A teacher demonstrates heating some naphthalene in a fume cupboard. The temperature of the naphthalene is measured every 2 minutes.

Figure 1

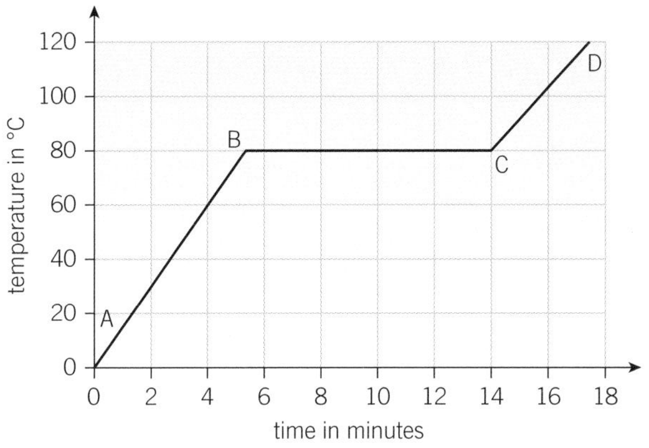

(Graph: temperature in °C (y-axis, 0 to 120) vs time in minutes (x-axis, 0 to 18). Points A at ~(1, 15) rising to B at ~(5, 80), flat line B to C at 80 °C from ~5 to ~14, then rising to D at ~(17, 120).)

02.1 Give the melting point of naphthalene. [1 mark]

02.2 Describe what is happening to the naphthalene between **A** to **B**, **B** to **C**, and **C** to **D**. [3 marks]

02.3 Suggest a reason why the teacher heated the naphthalene in a fume cupboard. [1 mark]

03.1 Particles in a solid are arranged in a regular fixed pattern. Draw diagrams to show the arrangement of particles in a liquid and a gas. [2 marks]

03.2 When a substance changes state, which property of the substance changes?
Choose the correct word from the box. [1 mark]

mass	temperature	volume

03.3 Calculate the energy required to melt an ice cube that has a mass of 0.075 kg.
The specific latent heat of fusion of ice is 3.34 × 10⁵ J/kg. [3 marks]

03.4 A bar tender prefers to use artificial, non-melting ice cubes to cool drinks. Estimate the final temperature of 200 g of water when the energy transferred from the artificial ice cube is 3360 joules. The water is at a temperature of 20 °C. Specific heat capacity of water is 4200 J/kg °C. [3 marks]

04 The apparatus in Figure 2 was used to measure the specific latent heat of vaporisation of water.

Figure 2

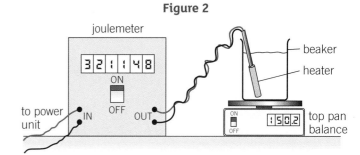

This is the method that was used.
1 Fill the beaker with water and switch on the heater.
2 When the water boils set the joulemeter to zero and take the reading on the top pan balance.
3 Allow the water to boil for 10 minutes.
4 Take the new readings on the joulemeter and top pan balance.
These are the results of the investigation.

Table 1

Time (mins)	Mass (g)	Energy (Joules)
0	184	0
10	168	37 800

04.1 Calculate the mass of water changed into steam in 10 minutes. [1 mark]

04.2 Calculate the latent heat of vaporisation of water using the results in **Table 1**.
Use the equation $E = mL$ [3 marks]

04.3 Give **one** reason why the value calculated in **04.2** is greater than the actual value. [1 mark]

04.4 Suggest **two** ways the investigation could be improved. [2 marks]

Learning objectives

After this topic, you should know:

- what a radioactive substance is
- the types of radiation given out from a radioactive substance
- when a radioactive source emits radiation (radioactive decay)
- there are different types of radiation emitted by radioactive sources.

A key discovery

If your photos showed a mysterious image, what would you think? In 1896 a French physicist, Henri Becquerel, discovered the image of a key on a photographic film he developed. He remembered the film had been in a drawer under a key. On top of that there had been a packet of uranium salts. The uranium salts must have sent out some form of radiation that passed through paper (the film wrapper) but not through metal (the key).

Becquerel asked a young physicist, Marie Curie, to investigate. She found that the salts emitted radiation all the time. She used the word radioactivity to describe this strange new property of uranium.

She and her husband, Pierre, did more research into this new branch of science. They discovered new radioactive elements. They named one of the elements polonium, after Marie's native country, Poland.

Investigating radioactivity

You can use a Geiger counter to detect radioactivity. This is made up of a detector called a Geiger–Müller tube (or Geiger tube) connected to an electronic counter (Figure 2). The counter clicks each time a particle of radiation from a radioactive substance enters the Geiger tube.

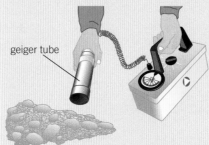

geiger tube

Figure 2 *Using a Geiger counter*

Safety: Avoid touching and inhaling radioactive material.

photographic plate

Figure 1 *Becquerel's key*

Inside the atom

What stops the radiation? The physicist Ernest Rutherford carried out tests to answer this question about a century ago. He put different materials between the radioactive substance and a detector.

He discovered two types of radiation:

- One type (**alpha radiation α**) was stopped by paper.
- The other type (**beta radiation β**) went through the paper.

Scientists later discovered a third type, **gamma radiation γ**, which is even more penetrating than beta radiation.

Rutherford carried out more investigations and discovered that alpha radiation is made up of positively charged particles. He realised that these particles could be used to probe the atom. His research students included Hans Geiger, who invented what was later called the Geiger counter. They carried out investigations in which a narrow beam of alpha particles was directed at a thin metal foil. Rutherford was astonished that some of the alpha particles rebounded from the foil. He proved that this happens because every atom has at its centre a positively charged nucleus containing most of the mass of the atom. He went on to propose that the nucleus contains two types of particle – protons and neutrons.

A radioactive puzzle

Why are some substances radioactive? Every atom has a nucleus made up of protons and neutrons. Electrons move about in energy levels (or shells) surrounding the nucleus.

Most atoms each have a stable nucleus that doesn't change. But the atoms of a radioactive substance each have a nucleus that is unstable. An unstable nucleus becomes stable or less unstable by emitting alpha, beta, or gamma radiation.

An unstable nucleus is described as decaying when it emits radiation. No one can tell exactly when an unstable nucleus will decay. It is a random event that happens without anything being done to the nucleus.

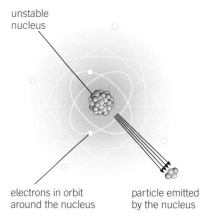

unstable nucleus

electrons in orbit around the nucleus

particle emitted by the nucleus

Figure 3 *Radioactive decay*

Synoptic link

For more about X-rays, look at Topic P12.5

1 a Write two differences between the radiation from uranium and the radiation from a lamp. [2 marks]
   b Write two differences between radioactive atoms compared with the atoms in a lamp filament. [2 marks]
2 a i The radiation from a radioactive source is stopped by paper. Name the type of radiation the source emits. [1 mark]
     ii The radiation from a different source goes through paper. Name the type of radiation this source emits. [1 mark]
   b Name the type of radiation from radioactive sources that is the most penetrating. [1 mark]
3 Explain why some substances are radioactive. [2 marks]
4 A Geiger counter clicks very rapidly when a certain substance is brought near it.
   a Describe the substance that made the Geiger counter click. [2 marks]
   b When the Geiger tube was near the substance, the counter clicked much less when a sheet of paper was placed between the substance and the tube. Explain why the counter clicked much less. [3 marks]

Key points

- A radioactive substance contains unstable nuclei that become stable by emitting radiation.
- There are three main types of radiation from radioactive substances – α, β, and γ.
- Radioactive decay is a random event – you can't predict or influence when it will happen.
- Radioactive sources emit α, β, and γ radiation.

P7.2 The discovery of the nucleus

Learning objectives

After this topic, you should know:

- how the nuclear model of the atom was established
- why the 'plum pudding' model of the atom was rejected
- what conclusions were made about the atom from experimental evidence
- why the nuclear model was accepted.

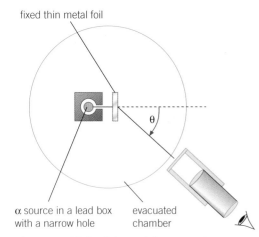

fixed thin metal foil

θ

α source in a lead box with a narrow hole

evacuated chamber

Figure 1 *Alpha (α) particle scattering*

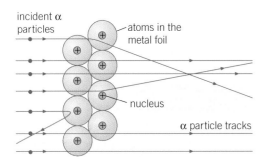

incident α particles

atoms in the metal foil

nucleus

α particle tracks

Figure 2 *Alpha (α) particle tracks*

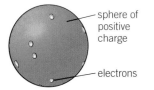

sphere of positive charge

electrons

Figure 3 *The plum pudding atom*

The physicist Ernest Rutherford discovered that alpha and beta radiation is made up of different types of particles. He realised that alpha (α) particles could be used to probe the atom. He asked two of his research workers, Hans Geiger and Ernest Marsden, to investigate how a thin metal foil scatters a beam of alpha particles. Figure 1 shows the arrangement they used.

The apparatus was in a vacuum chamber to prevent air molecules absorbing the alpha (α) particles. The detector consisted of a microscope focused on a small glass plate. Each time an alpha particle hit the plate, a spot of light was observed. The detector was moved to different positions. At each position, the number of spots of light observed in a certain time was counted.

Their results showed that:

- most of the alpha particles passed straight through the metal foil
- the number of alpha particles deflected per minute decreased as the angle of deflection increased
- about 1 in 10 000 alpha particles were deflected by more than 90°.

Rutherford was astonished by the results. He said it was like firing naval shells at tissue paper and discovering that a small number of the shells rebound. He knew that α particles are positively charged and that the radius of an atom is about 10^{-10} m. He deduced from the results that there is a positively charged nucleus at the centre of every atom that is:

- much smaller than the atom because most α particles pass through the atom without deflection
- where most of the mass of the atom is located.

Rutherford's nuclear model of the atom was quickly accepted because it:

- agreed exactly with the measurements Geiger and Marsden made in their experiments
- explained radioactivity in terms of changes that happen to an unstable nucleus when it emits radiation
- predicted the existence of the neutron, which was later discovered.

The plum pudding model

Before the nucleus was discovered in 1914, scientists didn't know what the structure of the atom was. They did know that atoms contain tiny negatively charged particles (which they called electrons).

Some scientists thought the atom was like a 'plum pudding' with positively charged matter evenly spread about (as in a pudding), and electrons buried inside (like plums in the pudding). But Rutherford's discovery meant that the plum pudding model of the atom was no longer accepted by scientists.

Bohr's model of the atom

After Rutherford's discovery, scientists knew that every atom has a positively charged nucleus that negatively charged electrons move around. The physicist Niels Bohr put forward the theory that the electrons in an atom orbit the nucleus at specific distances and specific energy values or energy levels. His model of the atom showed that the electrons in an orbit can move to another orbit by absorbing electromagnetic radiation to move away from the nucleus or by emitting electromagnetic radiation to move closer to the nucleus. His calculations based on his atomic model agreed with experimental observations of the light emitted by atoms.

A nuclear puzzle

More α-scattering experiments showed that:

- the hydrogen nucleus has the least amount of charge
- the charge of any nucleus is shared equally between a whole number of smaller particles, each with the same amount of positive charge.

The name proton was given to the hydrogen nucleus because scientists reckoned that every other nucleus contained hydrogen nuclei. But they also knew that the mass of every nucleus except for the hydrogen nucleus is bigger than the total mass of its protons. So there must be an uncharged type of particle with about the same mass as a proton in every nucleus except the hydrogen nucleus. They called this uncharged particle the neutron. The proton–neutron model of the nucleus explains all the mass and charge values of every nucleus. Direct experimental evidence for its existence was found by the physicist James Chadwick about 20 years after Rutherford's discovery of the nucleus.

Go further!

An atom emits electromagnetic radiation when an electron moving around the nucleus jumps from one energy level to a lower energy level. The radiation is emitted as a **photon**, which is a packet of waves emitted in a short burst. The energy of the emitted photon is equal to the energy change of the electron. Einstein put forward the photon theory and Bohr used it in his calculations.

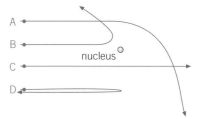

Figure 4 *See Summary question 2*

1 Write four features of every nucleus of every atom. [4 marks]

2 a Figure 4 shows four possible paths, labelled A, B, C, and D, of an alpha particle deflected by a nucleus. Choose the path the alpha particle would travel along. [1 mark]
  b Explain why each of the other paths in part **a** is not possible. [3 marks]

3 a i Write the conclusions that scientists made about the atom as a result of the discovery of electrons. [2 marks]
   ii Describe two differences between the nuclear model of the atom and the plum pudding model. [2 marks]
  b Explain why the alpha-scattering experiment led to the acceptance of the nuclear model of the atom and the rejection of the plum pudding model. [2 marks]

4 a Write one difference and one similarity between a proton and a neutron. ✓ [2 marks]
  b Explain why the mass of a helium nucleus is four times the mass of a hydrogen nucleus and its charge is only twice as much as the charge of a hydrogen nucleus. [3 marks]

Key points

- Rutherford used α particles to probe inside atoms. He found that some of the α particles were scattered through large angles.
- The 'plum pudding' model could not explain why some α particles were scattered through large angles.
- An atom has a small positively charged central nucleus where most of the atom's mass is located.
- The nuclear model of the atom correctly explained why some α particles scattered through large angles.

P7.3 Changes in the nucleus

Learning objectives

After this topic, you should know:

- what an isotope is
- how the nucleus of an atom changes when it emits an alpha particle or a beta particle
- how to represent the emission of an alpha particle from a nucleus
- how to represent the emission of a beta particle from a nucleus.

Table 1 *Relative mass and charge of subatomic particles*

	Relative mass	Relative charge
proton	1	+1
neutron	1	0
electron	$\sim\dfrac{1}{2000}$	−1

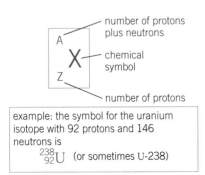

example: the symbol for the uranium isotope with 92 protons and 146 neutrons is

$^{238}_{92}U$ (or sometimes U-238)

Figure 1 *Representing an isotope*

the nucleus emits an α particle and forms a new nucleus

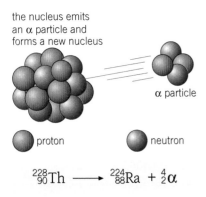

α particle

proton   neutron

$^{228}_{90}Th \longrightarrow {}^{224}_{88}Ra + {}^{4}_{2}\alpha$

Figure 2 α *emission*

In alpha (α) or beta (β) decay, the number of protons in a nucleus changes. In α decay, the number of neutrons also changes. How do the changes happen in α and β decay, and how can you represent these changes?

Table 1 gives the relative masses and the relative electric charges of a proton, a neutron, and an electron.

The **atomic number** (or proton number) of a nucleus is the number of protons in it. It has the symbol Z. Atoms of the same element each have the same number of protons.

The **mass number** of a nucleus is the number of protons plus neutrons in it. It has the symbol A.

Isotopes are atoms of the same element with different numbers of neutrons. The isotopes of an element have nuclei with the same number of protons but a different number of neutrons. Figure 1 shows how to represent an isotope of an element X, which has Z protons and A protons plus neutrons.

Radioactive decay

An unstable nucleus becomes more stable by emitting an α (alpha) or a β (beta) particle or by emitting a γ (gamma) ray.

α emission

An α particle is made up of two protons plus two neutrons. Its relative mass is 4, and its relative charge is +2. So it is usually represented by the symbol $^{4}_{2}\alpha$. It is identical to a helium nucleus, so in nuclear equations you might see it represented by the symbol $^{4}_{2}He$.

When an unstable nucleus emits an α particle:

- its atomic number goes down by 2, and its mass number goes down by 4
- the mass and the charge of the nucleus are both reduced.

For example, the thorium isotope $^{228}_{90}Th$ decays by emitting an α particle. So, it forms the radium isotope $^{224}_{88}Ra$.

Figure 2 shows an equation to represent this decay.

- The numbers along the top show that the total number of protons and neutrons after the change (= 224 + 4) is equal to the total number of neutrons and protons before the change (= 228).
- The numbers along the bottom show that the total number of protons after the change (= 88 + 2) is equal to the total number of protons before the change (= 90).

β emission

A β particle is an electron created and emitted by a nucleus that has too many neutrons compared with its protons. A neutron in the nucleus changes into a proton and a β particle (i.e. an electron), which is instantly emitted. The relative mass of a β particle is effectively zero, and its relative charge is −1. So a β particle can be represented by the symbol $^{0}_{-1}\beta$.

When an unstable nucleus emits a β particle:

● the atomic number of the nucleus goes up by 1, and its mass number is unchanged (because a neutron changes into a proton)

● the charge of the nucleus is increased, and the mass of the nucleus is unchanged.

For example, the potassium isotope $^{40}_{19}K$ decays by emitting a β particle. So it forms a nucleus of the calcium isotope $^{40}_{20}Ca$. Figure 3 shows an equation to represent this decay.

● The numbers along the top show that the total number of protons and neutrons after the change (= 40 + 0) is equal to the total number of neutrons and protons before the change (= 40).

● The numbers along the bottom show that the total charge (in relative units) after the change (= 20 − 1) is equal to the total charge before the change (= 19).

γ emission

A γ-ray is electromagnetic radiation from the nucleus of an atom. It is uncharged and has no mass. So its emission does not change the number of protons or neutrons in a nucleus. So the mass and the charge of the nucleus are both unchanged.

Neutron emission

Neutrons are emitted by some radioactive substances as a result of α particles colliding with unstable nuclei in the substance. Such a collision causes the unstable nuclei to become even more unstable and emit a neutron. Because the emitted neutrons are uncharged, they can pass through substances more easily than an α particle or a β particle can.

1 How many protons and how many neutrons are there in the nucleus of each of the following isotopes:
   a $^{12}_{6}C$ [1 mark]   **b** $^{60}_{27}Co$ [1 mark]   **c** $^{235}_{92}U$? [1 mark]
   d How many more protons and how many more neutrons are in $^{238}_{92}U$ compared with $^{224}_{88}Ra$? [2 marks]

2 A substance contains the radioactive isotope $^{238}_{92}U$, which emits alpha radiation. The product nucleus X emits beta radiation and forms a nucleus Y. Determine how many protons and how many neutrons are present in:
   a a nucleus of $^{238}_{92}U$ [1 mark]   **b** a nucleus of X [2 marks]
   c a nucleus of Y [2 marks]

3 Copy and complete the following equations for α and β decay.
   a $^{238}_{92}U \rightarrow {}^{?}_{?}Th + {}^{4}_{2}\alpha$ **b** $^{64}_{29}Cu \rightarrow {}^{?}_{?}Zn + {}^{0}_{-1}\beta$ [4 marks]

4 A radioactive isotope of polonium (Po) has 84 protons and 126 neutrons. The isotope is formed from the decay of a radioactive isotope of bismuth, which emits a β particle in the process. Copy and complete the equation below to represent this decay.

   $$Bi \rightarrow Po + \beta$$ [3 marks]

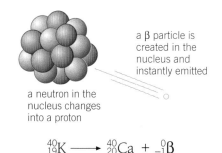

a β particle is created in the nucleus and instantly emitted

a neutron in the nucleus changes into a proton

$$^{40}_{19}K \longrightarrow {}^{40}_{20}Ca + {}^{0}_{-1}\beta$$

Figure 3 β emission

Go further!

Most nuclei are stable because the protons and neutrons inside a nucleus are held together by a strong attractive force called the strong nuclear force. This force is strong enough in stable nuclei to overcome the electrostatic repulsion between protons, and to stop the neutrons moving away from the nucleus.

Study tip

Make sure you know the changes to mass number and to atomic number in alpha decay and in beta decay.

Key points

● Isotopes of an element are atoms with the same number of protons but different numbers of neutrons. So they have the same atomic number but different mass numbers.

α decay	β decay
Change in the nucleus	
Nucleus loses 2 protons and 2 neutrons	A neutron in the nucleus changes into a proton
Particle emitted	
2 protons and 2 neutrons emitted as an α particle	An electron is created in the nucleus and instantly emitted
Equation	
$^{A}_{Z}X \rightarrow {}^{A-4}_{Z-2}Y + {}^{4}_{2}\alpha$	$^{A}_{Z}X \rightarrow {}^{A}_{Z+1}Y + {}^{0}_{-1}\beta$

P7.4 More about alpha, beta, and gamma radiation

Learning objectives

After this topic, you should know:

- how far each type of radiation can travel in air
- how different materials absorb alpha, beta, and gamma radiation
- the ionising power of alpha, beta, and gamma radiation
- why alpha, beta, and gamma radiation is dangerous.

Table 1 *The results of the two tests*

Type of radiation	Absorber materials	Range in air
alpha α	Thin sheet of paper	about 5 cm
beta β	Aluminium sheet (about 5 mm thick) Lead sheet (2–3 mm thick)	about 1 m
gamma γ	Thick lead sheet (several cm thick) Concrete (more than 1 m thick)	unlimited – spreads out in air without being absorbed

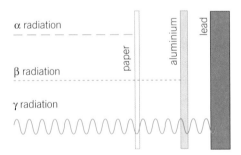

Figure 2 *The penetrating power of alpha, beta, and gamma radiation*

Penetrating power

Alpha radiation can't penetrate paper. But what stops beta and gamma radiation? And how far can each type of radiation travel through air? You can use a Geiger counter to find out, but you must take account of background radiation, which is radiation from unstable nuclei in materials around us and in the atmosphere. To do this you need to:

1. measure the count rate (which is the number of counts per second) without the radioactive source present. This is the background count rate.

2. measure the count rate with the source in place. Subtracting the background count rate from this gives you the count rate from the source alone.

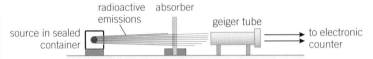

Figure 1 *Absorption tests*

You can then test absorber materials and the range that each type of radiation travels in air.

To test different materials, you need to place each material between the tube and the radioactive source (Figure 1). Then you measure the count rate. You can add more layers of material until the count rate from the source is zero. The radiation from the source has then been stopped by the absorber material.

To test the range that each type of radiation travels in air, you need to move the tube away from the source. When the tube is beyond the range of the radiation, the count rate from the source is zero.

Radioactivity dangers

The radiation from a radioactive substance can knock electrons out of atoms. The atoms become charged because they lose electrons. The process is called **ionisation**. When an object is exposed to ionising radiation, it is said to be **irradiated**, but it does not become radioactive.

Radioactive substances can contaminate other materials that they come into contact with. **Radioactive contamination** is the unwanted presence of materials containing radioactive atoms on other materials. The hazard from contamination is due to the decay of the nuclei of the contaminating atoms. The type of radiation emitted affects the level of hazard.

X-rays, fast-moving protons, and fast-moving neutrons also cause ionisation. Ionisation in a living cell can damage or kill the cell. Damage to the genes in a cell can be passed on if the cell generates more cells. Strict safety rules must always be followed when radioactive substances are used.

Alpha radiation is more dangerous in the body than beta or gamma radiation. This is because the ionising power of alpha radiation is much greater than the ionising power of beta or gamma radiation.

Workers who use ionising radiation reduce their exposure by:

● keeping as far away as possible from the source of radiation (e.g., by using special handling tools with long handles)

● spending as little time as possible in at-risk areas

● shielding themselves by staying behind thick concrete barriers and/or using thick lead plates.

Peer review

Scientists have studied the effects of radiation on humans, including the survivors of the atom bombs dropped on Japan in 1945. Their findings are published and shared with other scientists so that the findings can be checked by them. This process is called peer review.

Radiation in use

When a radioactive substance is used, the substance must emit the appropriate type of radiation for that use.

Smoke alarms contain a radioactive isotope that sends out alpha particles into a gap in a circuit in the alarm. The alpha particles ionise the air in the gap so there is a current across the gap. In a fire, smoke absorbs the alpha particles, preventing them from ionising the air, so the current across the gap drops and the alarm sounds. Beta or gamma radiation could not be used because they do not create enough ions to make the air in the gap conduct electricity.

Automatic thickness monitoring in metal foil production uses a radioactive source that sends out β radiation (Figure 3). The amount of β radiation passing through the foil depends on the thickness of the foil. The detector measures the amount of radiation passing through the foil. If the foil is too thick, the detector reading drops and the detector sends a signal to increase the pressure of the rollers on the metal sheet. This makes the foil thinner again. Gamma radiation isn't used because it would all pass through the foil unaffected. Alpha radiation isn't used as it would all be stopped by the foil.

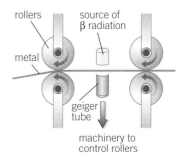

Figure 3 *Thickness monitoring using a radioactive source*

> ## Key points
>
> ● α radiation is stopped by paper and has a range of a few centimetres in air. It consists of particles, each composed of two protons and two neutrons. It has the greatest ionising power.
>
> ● β radiation is stopped by a thin sheet of metal and has a range of about one metre in air. It consists of fast-moving electrons emitted from the nucleus. It is less ionising than alpha radiation and more ionising than gamma radiation.
>
> ● γ radiation is stopped by thick lead and has an unlimited range in air. It consists of electromagnetic radiation.
>
> ● Alpha, beta, and gamma radiation ionise substances they pass through. Ionisation in a living cell can damage or kill the cell.

1 a Explain why a radioactive source is stored in a lead-lined box. [1 mark]

 b Name the type of ionising radiation from radioactive substances that is most easily absorbed. [1 mark]

 c Name the type or types of radiation from a radioactive source that are stopped by a thick aluminium plate. [1 mark]

2 a Name the type of radiation from a radioactive source that is:

  i uncharged [1 mark]

  ii positively charged [1 mark]

  iii negatively charged. [1 mark]

 b Name the type of radiation from a radioactive source that:

  i has the longest range in air [1 mark]

  ii has the greatest ionising power. [1 mark]

3 a Explain why ionising radiation is dangerous. [2 marks]

 b Explain how you would use a Geiger counter to find the range of the radiation from a source of α radiation. [3 marks]

4 Explain why γ radiation is not suitable for monitoring the thickness of metal foil. [2 marks]

P7.5 Activity and half-life

Learning objectives

After this topic, you should know:

- what is meant by the half-life of a radioactive source
- what is meant by the count rate from a radioactive source
- what happens to the count rate from a radioactive isotope as it decays
- how to calculate count rates after a given number of half-lives. **H**

Half-life calculations

Figure 1 shows that the count rate from a sample of a radioactive isotope decreases from 600 to 300 to 150 to 75 after three successive half-life intervals. In general, you can work out the count rate or the number of unstable nuclei left after n half-lives by dividing the initial value by 2 to the power n (i.e., 2 multiplied by itself n times). You can write this as an equation:

$$\text{count rate (number of unstable nuclei) after } n \text{ half-lives} = \frac{\text{initial count rate (number of unstable nuclei)}}{2^n}$$

Worked example

A particular radioactive isotope has a half-life of 6.0 hours. A sample of this isotope contains 60 000 radioactive nuclei. Calculate the number of radioactive nuclei of this isotope remaining after 24 hours.

Solution

$n = 4$ because 24 hours equals 4 half-lives for this isotope.

$2^4 = 16$, so the number of radioactive nuclei of the isotope remaining after 24 hours = $60\,000 \div 2^4 = 60\,000 \div 16$ = **3750**

Every atom of an element always has the same number of protons in its nucleus. But the number of neutrons in the nucleus can differ. An atom of a specific element with a certain number of neutrons is called an isotope of that element.

The **activity** of a radioactive source is the number of unstable atoms in the source that decay per second. The unit of activity is the Becquerel (Bq), which is 1 decay per second. As the nucleus of each unstable atom (the parent atom) decays, the number of parent atoms decreases. So the activity of the sample decreases.

You can use a Geiger counter to monitor the activity of a radioactive sample. To do this, you need to measure the **count rate** from the sample. The count rate is the number of counts per second. This is proportional to the activity of the source, as long as the distance between the tube and the source stays the same. The graph in Figure 1 shows that the count rate of a sample decreases with time.

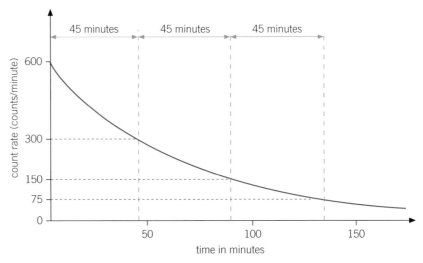

Figure 1 *A graph of count rate against time. The count rate here is measured in counts per minute*

The average time taken for the count rate (and so the number of parent atoms) to fall by half is always the same. This time is called the **half-life**. The half-life shown on the graph is 45 minutes.

The half-life of a radioactive isotope is the average time it takes:

- for the number of nuclei of the isotope in a sample (and so the mass of parent atoms) to halve
- for the count rate from the isotope in a sample to fall to half its initial value.

The random nature of radioactive decay

Radioactive decay is a random process. This means that no one can predict exactly *when* an individual atom will suddenly decay. But you *can* predict how many atoms will decay in a given time – because there are so many of them. This is a bit like throwing dice. You can't predict what number you will get with a single throw. But if you threw 1000 dice, you would expect one-sixth to come up with a particular number.

Suppose you start with 1000 unstable atoms. Look at the graph in Figure 2.

If 10% decay every hour:

- 100 atoms will decay in the first hour, leaving 900 atoms
- 90 atoms (= 10% of 900) will decay in the second hour, leaving 810 atoms.

Table 1 shows what you get if you continue the above calculations. The results are plotted as a graph in Figure 2. The graph is like Figure 1, except the half life is just over 6 hours. The similarity is because radioactive decay, like throwing dice, is a random process.

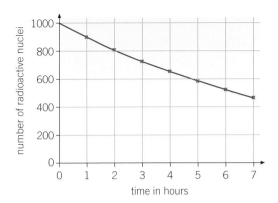

Figure 2 *Half-life*

Table 1 *What you get if you continue the calculations. The results are plotted as a graph in Figure 2*

Time from start (in hours)	No. of unstable atoms present	No. of unstable atoms that decay in the next hour
0	1000	100
1	900	90
2	810	81
3	729	73
4	656	66
5	590	59
6	531	53
7	478	48

1   a   Define the half-life of a radioactive isotope. [1 mark]
     b   Determine what the count rate in Figure 1 will be after 75 minutes from the start. [1 mark]

2   A radioactive isotope has a half-life of 15 hours. A sealed tube contains 8 milligrams of this isotope.
     a   Calculate what mass of the isotope is in the tube:
         i   15 hours later [1 mark]
         ii 45 hours later. [1 mark]
     b   Estimate how long it would take for the mass of the isotope to decrease to less than 5% of the initial mass. [3 marks]

3   a   Ⓗ A sample of a radioactive isotope contains 320 million atoms of the isotope.
         i   Calculate how many atoms of the isotope are present after one half-life. [1 mark]
         ii Calculate the ratio of the number of atoms of the isotope left after five half lives to the initial number of atoms. [1 mark]
         iii Calculate the number of atoms of the isotope left after five half-lives. [2 marks]
     b   Estimate how long it would take for the count rate in Figure 1 to decrease to less than 40 counts per minute. [2 marks]

4   A sample of old wood was carbon dated and found to have 25% of the count rate measured in an equal mass of living wood. The half-life of the radioactive carbon is 5600 years. Calculate the age of the sample of wood. [2 marks]

Key points

- The half-life of a radioactive isotope is the average time it takes for the number of nuclei of the isotope in a sample to halve.
- The count rate of a Geiger counter caused by a radioactive source decreases as the activity of the source decreases.
- The number of atoms of a radioactive isotope and the count rate both decrease by half every half-life.
- Ⓗ The count rate after n half-lives = the initial count rate ÷ 2^n.

P7 Radioactivity

Summary questions

1 a Calculate how many protons and how many neutrons are in a nucleus of each of the following isotopes:

      i $^{14}_{6}C$ ii $^{228}_{90}Th$.   [2 marks]

   b $^{14}_{6}C$ emits a **β** particle and becomes an isotope of nitrogen (N).

      i Write how many protons and how many neutrons are in this nitrogen isotope. [2 marks]

      ii Write the symbol for this isotope.   [1 mark]

   c $^{228}_{90}Th$ emits an **α** particle and becomes an isotope of radium (Ra).

      i Write how many protons and how many neutrons are in this isotope of radium. [2 marks]

      ii Write the symbol for this isotope.   [1 mark]

2 Copy and complete the following table about the properties of alpha, beta, and gamma radiation.   [4 marks]

	α	β	γ
Identity		electrons	
Stopped by			thick lead
Range in air		about 1 m	
Relative ionisation			weak

3 The following measurements were made of the count rate from a radioactive source.

Time in hours	0	0.5	1.0	1.5	2.0	2.5
Count rate due to the source in counts per minute	510	414	337	276	227	188

   a Plot a graph of the count rate (on the vertical axis) against time.   [3 marks]

   b Use your graph to find the half-life of the source.   [1 mark]

4 In a radioactive carbon dating experiment of ancient wood, a sample of the wood had an activity of 40 Bq. The same mass of living wood had an activity of 320 Bq.

   a i Explain what is meant by the activity of a radioactive source.   [1 mark]

      ii **H** Calculate how many half-lives the activity took to decrease from 320 to 40 Bq.   [2 marks]

   b The half-life of the radioactive carbon in the wood is 5600 years. Calculate the age of the sample.   [1 mark]

5 In an investigation to find out what type of radiation was emitted from a given source, the following measurements were made with a Geiger counter.

Source at 20 mm from tube	Average count rate in counts per minute
no source present	29
no absorber present	385
sheet of metal foil between S and T	384
thick aluminium plate between S and T	32

   a Write what caused the count rate when no source was present.   [1 mark]

   b Write the count rate from the source with no absorbers present.   [1 mark]

   c Write what type of radiation was emitted by the source. Explain how you arrived at your answer.   [4 marks]

6 Figure 1 shows the path of two **α** particles labelled A and B that are deflected by the nucleus of an atom.

Figure 1

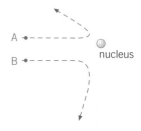

   a Explain why they are deflected by the nucleus. [2 marks]

   b Explain why B is deflected less than A.   [2 marks]

   c Explain why most **α** particles directed at a thin metal foil pass straight through it.   [2 marks]

7 a Explain what is meant by the activity of a radioactive source ?   [1 mark]

   b The radioactive isotope $^{220}_{82}Rn$ emits alpha particles to form an isotope of polonium (Po). Write down the equation for this process.   [2 marks]

   c $^{220}_{82}Rn$ has a half-life of 52 s. A pure sample of the isotope has an initial activity of 800 Bq. What is its activity after **i** 104 s, and **ii** 260 s?   [4 marks]

Practice questions

01 A group of students investigated the nature of radioactive decay. They used 80 one-penny coins and a stopwatch to perform the investigation.

1 Place the 80 one-penny coins in a container with lid.

2 Shake container for 10 seconds and tip the coins onto the bench.

3 Count the number of coins with the heads side showing and record result.

4 Replace only the coins showing the tails side into container and repeat the tests.

Table 1

Time in seconds	Number of heads shown
0	80
10	39
20	20
30	11
40	5
50	3
60	2
70	1

01.1 Draw a graph of number of heads (y-axis) against time (x-axis). [3 marks]

01.2 Give some conclusions about the results. [2 marks]

01.3 Give a reason why only one-penny coins were used and not a mix of coins. [1 mark]

01.4 Suggest one reason for a possible human error in the investigation. [1 mark]

01.5 Explain what the head coins are meant to represent in nuclear decay. [1 mark]

02 The physicist Ernest Rutherford and two of his research workers, Geiger and Marsden, investigated how a thin metal foil scattered a beam of alpha particles. Their results showed that:

● most of the alpha particles passed straight through the metal foil

● the number of alpha particles deflected per minute decreased as the angle of deflection increased

● about 1 in 10 000 alpha particles were deflected by more than 90°.

02.1 Explain how the Rutherford investigation led to the Plum Pudding model of an atom being replaced. [3 marks]

02.2 Describe how the new proposed model of an atom was further changed by the work of the physicist Niels Bohr [2 marks]

03.1 Complete the table of information for a radon atom.

Number of protons	86
Number of electrons	
Number of neutrons	132

[1 mark]

03.2 Calculate the mass number of this radon atom. [1 mark]

03.3 Radon can exist in the form of many different isotopes. Describe the differences in the nuclei of these isotopes. [1 mark]

03.4 The half-life of radon-222 is 3.8 days. Calculate how long a 48 g sample of radon will last before it contains only 3 g of radon and stops being effective. [2 marks]

04.1 Match each type of radiation to the correct description.

alpha radiation	electromagnetic radiation
beta radiation	same as a helium nucleus
gamma radiation	high-speed electron

[3 marks]

04.2 When nuclear isotopes decay nuclear radiation is emitted. Complete the nuclear decay diagram.

$$^{230}_{90}\text{Th} \xrightarrow{\alpha} {}^{226}\text{Ra} \longrightarrow {}_{86}\text{Rn}$$ [3 marks]

A sample of radioactive material is tested to find out whether it passes through different materials. The results are shown in **Figure 1**.

Figure 1

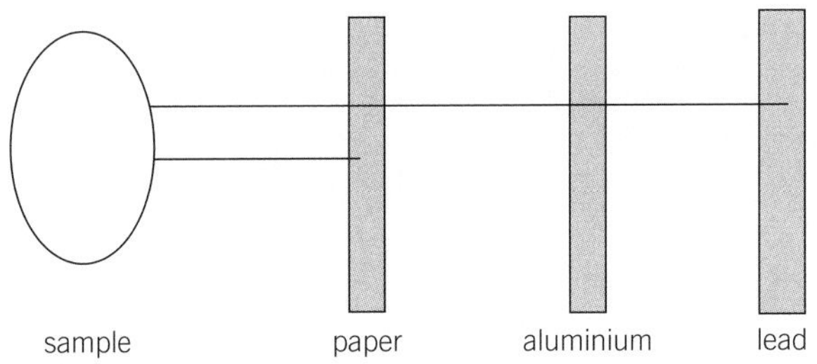

sample   paper   aluminium   lead

04.3 Name the types of radiation in the sample. [1 mark]

3 Forces in action

An astronaut in a space station can float around and perform acrobatic tricks. Many people think this is because there is no gravity in space. However, this is wrong. The force of gravity due to the Earth stretches far into space and keeps the space station orbiting the Earth.

You and all of the objects around you are acted on by the force of gravity. You are also acted on by other forces, such as friction, which acts between objects when they touch each other, and non-contact forces like magnetic and electrostatic forces.

In this section, you will learn about what forces do, how we measure them and their effects, and how we calculate the effect forces have on objects.

Key questions

- How do we represent a force and what do we mean by a resultant force?

- How do we work out the effect of a resultant force acting on an object?

- What do we mean by momentum?

- What do we mean by elasticity?

Making connections

- When a force does work on an object to make an it move faster, energy is transferred to the object to increase its kinetic energy store. As you work through this section you will meet content that describes what happens when forces do work on objects. Make sure that you look back at **P1 Conservation and dissipation of energy** to recall the different types of energy transfers between objects.

I already know...

Force is measured in newtons (N) using a newton-meter.

An object is in equilibrium when the forces acting on it are balanced.

The weight of an object is due to the force of gravity on it.

Speed is measured in metres per second.

Drag forces and friction resist the motion of moving objects.

When objects interact, each one exerts a force on the other.

The force in a stretched object is called tension and it increases if the object is stretched more.

I will learn...

The difference between a vector and a scalar and how to represent a vector.

How to find the resultant of two forces and to resolve a force into perpendicular components.

How to calculate the weight of an object from its mass and the gravitational field strength of where it is.

The difference between speed and velocity and what we mean by acceleration.

What is meant by terminal velocity and why objects fall through water at a constant velocity.

What is meant by conservation of momentum and when we can use this rule.

How to measure the stiffness of a spring and what is meant by elasticity.

Required Practicals

Practical		Topic
18	Stretch tests	P10.5
19	Investigating force and acceleration	P10.1

Learning objectives

After this topic, you should know:

- what do we mean by displacement
- what is meant by a vector quantity
- what is meant by a scalar quantity
- how to represent a vector quantity.

Distance and displacement

When you travel to school, the distance you travel may be much greater than the direct distance from your home to your school. The map in Figure 1 shows the route from home to school for a student who has to catch a bus to school. The bus has to pick up lots of other students along a route that has quite a few changes of direction. So the distance travelled by the student is much greater than the direct distance from the student's home to the school.

Distance without change of direction is called **displacement**. In other words, displacement is distance in a certain direction.

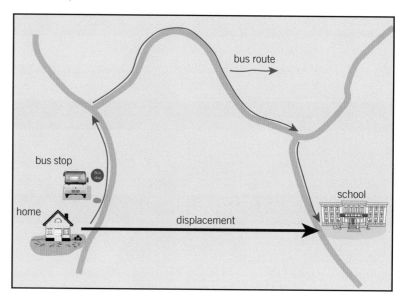

Figure 1 *A school journey*

Vectors and scalars

As well as velocity and displacement, many other physical quantities have both a *size* and a *direction*. Physical quantities that have direction are called **vectors**. Other examples of vectors include acceleration, force, momentum, weight, and gravitational field strength.

Physical quantities that have size, but no specific direction are called **scalars**. Examples include speed, distance, time, mass, energy, and power.

The size of a quantity is its **magnitude**. A vector has magnitude (i.e., size) as well as a direction. A scalar has magnitude only. For example, in Figure 1 the displacement from home to school is 5 km due East. So the magnitude of the displacement is 5 km, and the direction is East.

- A vector quantity has a magnitude and a direction.
- A scalar quantity has magnitude only.

Representing a vector quantity

Any vector quantity can be represented by an arrow, like the displacement arrow in Figure 1.

- The direction of the arrow shows the direction of the vector quantity.
- The length of the arrow represents the magnitude of the vector quantity.

Because a force has a magnitude and a direction, it is a vector quantity and can be represented on a diagram by an arrow. Figure 2 shows the force acting on a nail when it is struck by a hammer. The force of the hammer on the nail is represented by the red arrow (Figure 2). If the magnitude of the force is known, this can also be represented by the length of the arrow.

Scale diagrams

When more than one force acts on an object, the forces on the object sometimes need to be shown on a scale diagram. For example, suppose two forces of 3.0 N and 4.0 N act at right angles to each other on a small object. To show both forces on a scale diagram, we could choose a scale in which 1 unit of distance (for example 10 mm) represents a force of 1.0 N. On this scale, the length of the two arrows would need to be 30 mm and 40 mm respectively (Figure 3).

1 a Explain what is meant by the magnitude of a vector quantity. [1 mark]

b Describe the difference between a scalar quantity and a vector quantity. [1 mark]

2 Look at the journey shown on the map in Figure 1. If the displacement arrow represents a displacement of 10 km, use the map to estimate the approximate distance travelled by the student on their journey to school. [1 mark]

3 A small object is acted on by a horizontal force, **A**, of magnitude 15 N, and another horizontal force, **B**, of magnitude 12 N, which acts in the opposite direction to A. Draw a to-scale vector diagram showing forces **A** and **B** acting on the object. [2 marks]

4 A force **A** of 48 N acts on a small object as shown in Figure 4.

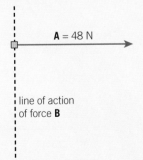

Figure 4

a Copy Figure 4 and write the scale you have used on your diagram. [1 mark]

b Add a further arrow to your diagram to represent a force **B** of 36 N acting on the object in a direction at right angles to the direction of the 48 N force. [1 mark]

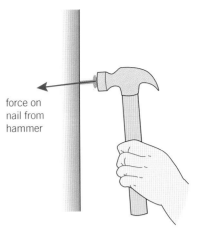

Figure 2 *Representing a force*

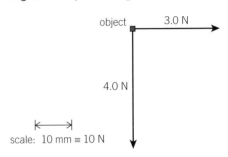

scale: 10 mm ≡ 10 N

Figure 3 *A scale diagram*

Study tip

Force diagrams usually show more than one force. For this reason, the forces on a diagram should always be clearly labelled to identify the force. Force diagrams do not always need to be scale diagrams, but if a diagram is a scale diagram, the scale should be shown (e.g., 10 mm ≡ 1.0 N, where ≡ means 'represents').

Key points

- Displacement is distance in a given direction.
- A vector quantity is a physical quantity which has magnitude and direction.
- A scalar quantity has magnitude but no direction.
- A vector quantity can be represented by an arrow in the direction of the vector and of length in proportion to the magnitude of the vector.

P8.2 Forces between objects

Learning objectives

After this topic, you should know:

- what forces can do
- the unit of force
- what is meant by a contact force
- the forces being exerted when two objects interact.

When you apply **forces** to a tube of toothpaste, the forces you apply squeeze the tube and change its shape and push toothpaste out of the tube. Be careful not to apply too much force, if you do the toothpaste might come out too fast. Forces can change the shape of an object or change its state of rest or its motion.

A force is a push or pull that acts on an object because of its interaction with another object. If two objects must touch each other to interact, the forces are called contact forces. Examples include friction, air resistance, stretching forces (or tension), and normal contact forces. Contact forces occur when an object is supported by or strikes another object. Non-contact forces include magnetic force, electrostatic force, and the force of gravity.

Synoptic link

You will meet Newton's second law when you study forces and acceleration in Topic P10.1.

Equal and opposite forces

Newton's third law of motion states that when two objects interact with each other, they exert equal and opposite forces on each other. The unit of force is the newton (abbreviated as N).

- A boxer who punches a bag with a force of 100 N experiences an equal and opposite force of 100 N from the bag.

- The weight of an object is the force of gravity on the object due to the Earth. The object exerts an equal and opposite force on the Earth.

- Two roller skaters pull on opposite ends of a rope (Figure 1). The skaters move towards each other because they pull on each other with equal and opposite forces. Two newton-meters could be used to show this.

Study tip

Remember that when two objects interact, although they exert equal and opposite forces on each other, the effects of these forces on each object will depend on the masses of the objects – the larger the mass, the smaller the effect.

Figure 1 *Equal and opposite forces*

In the mud

A car stuck in mud can be difficult to move. Figure 2 shows how a tractor can be very useful here. At any stage, the force of the rope on the car is equal and opposite to the force of the car on the rope.

To pull the car out of the mud, the force of the mud on the tractor needs to be greater than the force of the mud on the car. These two forces are not equal and opposite to each other. The 'equal and opposite force' to the force of the mud on the tractor is the force of the tractor on the mud. The 'equal and opposite force' to the force of the mud on the car is the force of the car on the mud.

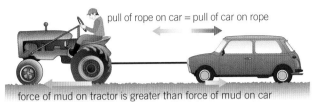

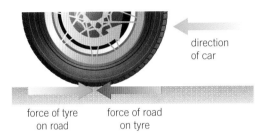

pull of rope on car = pull of car on rope

force of mud on tractor is greater than force of mud on car

Figure 2 *In the mud*

Friction in action

The **driving force** on a car is the force that makes it move. This is sometimes called the engine force or the motive force. This force pushes the car forward because there is **friction** between the ground and the tyre of each drive wheel. Friction acts where the tyre is in contact with the ground.

When the car moves forward:

● the force of the friction of the road on the tyre is in the forward direction

● the force of the friction of the tyre on the road is in the reverse direction.

These two forces are equal and opposite to each other (Figure 3).

direction of car

force of tyre   force of road
on road   on tyre

Figure 3 *Driving force*

1 **a** The brakes of a moving car are applied. Describe the effect of the braking force on the car. [1 mark]

   b When a car brakes, the road exerts a force on each tyre to slow the car down. Describe the force that each tyre exerts on the road. [1 mark]

2 **a** A hammer hits a nail with a downward force of 50 N. What is the size and direction of the force of the nail on the hammer? [1 mark]

   b A lorry tows a broken-down car. The force of the lorry on the tow rope is 200 N. How much force is exerted on the tow rope by the lorry? [1 mark]

3 A book is at rest on a table. Compare the force of the book on the table with each of the following forces:
   a the force of the table on the book [2 marks]
   b the force of the table on the floor. [2 marks]

4 When a student is standing at rest on bathroom scales, the scales read 500 N.
   a What is the size and direction of the force of the student on the scales? [1 mark]
   b What is the size and direction of the force of the scales on the student? [1 mark]
   c What is the size and direction of the force of the floor on the scales? [1 mark]

Key points

● Forces can change the shape of an object, or change its motion or its state of rest.
● The unit of force is the newton (N).
● A contact force is a force that acts on objects only when the objects touch each other.
● When two objects interact, they always exert equal and opposite forces on each other.

P8.3 Resultant forces

Learning objectives

After this topic, you should know:

- what a resultant force is
- what happens if the resultant force on an object is:
  - zero
  - greater than zero
- how to calculate the resultant force when an object is acted on by two forces acting along the same line
- what a free-body force diagram is. **H**

Figure 1 *The linear air track*

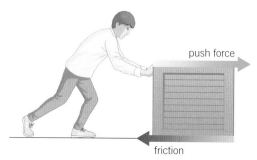

Figure 2 *Overcoming friction*

Wherever you are right now, at least two forces are acting on you. These are the gravitational force on you and a force supporting you. Most objects around you are acted on by more than one force. You can work out the effect of the forces on an object by replacing them with a single force, the **resultant force**. This is a single force that has the same effect as all the forces acting on the object. If the resultant force is zero, we say that the forces acting on the object are balanced.

Balanced forces

Newton's first law of motion states that if the forces acting on an object are balanced, the resultant force on the object is zero, and:

- if the object is at rest, it stays stationary
- if the object is moving, it keeps moving with the same speed and in the same direction.

If only two forces act on an object with zero resultant force, the forces must be equal to each other and act in opposite directions.

1. A glider on a linear air track floats on a cushion of air (Figure 1). As long as the track stays level, the glider moves at the same speed and direction along the track. That is because friction is absent. Newton's First law tells you that the glider will continue moving with the same speed in the same direction.

2. When a heavy crate is pushed across a rough floor at a constant speed without changing its direction, the push force on it is equal in size, and acting in the opposite direction, to the friction of the floor on the crate (Figure 2). Newton's first law states that the crate will continue moving with the same speed, and in the same direction.

Unbalanced forces

When the resultant force on an object is not zero, the forces acting on the object are not balanced. The movement of the object depends on the size and direction of the resultant force.

1. When a jet plane is taking off, the thrust force of its engines is greater than the force of air resistance (or drag) on it. The resultant force on the plane is the difference between the thrust force and the force of air resistance acting on it. The resultant force is therefore greater than zero.

2. When a car driver applies the brakes, the braking force is greater than the force from the engine. The resultant force is the difference between the braking force and the engine force. It acts in the opposite direction to the car's direction, so it slows the car down.

The examples show that if an object is acted on by two unequal forces acting in opposite directions, the resultant force is:

- equal to the difference between the two forces
- in the direction of the larger force.

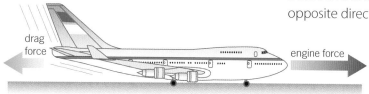

Figure 3 *A passenger jet on take-off*

For example, Figure 4 shows two forces, **A** and **B**, acting on an object in opposite directions. If **A** = 5 N and **B** = 9 N, the resultant force on the object is 4 N (= 9 N − 5 N) in the direction of B. If the two forces act in the same direction, the resultant force is equal to the sum of the two forces and is in the same direction.

Figure 4 *Forces in opposite directions*

Figure 5 shows a tug-of-war in which the pull force of each team is represented by a vector. A scale of 10 mm to 200 N is used. Team A pulls with a force of 1000 N, and team B pulls with a force of 800 N. So the resultant force is 200 N in team A's direction.

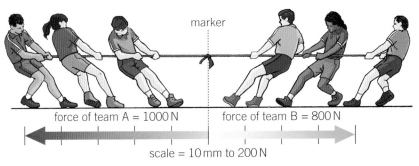

Figure 5 *A tug-of-war*

 Force diagrams

When an object is acted on by more than one force, you can draw a free-body force diagram to work out the resultant force on the object. A **free-body force diagram** shows the forces acting on an object without any other objects or other forces shown. Each force is shown on the diagram by a vector, which is an arrow pointing in the direction of the force. Figure 4 is a simple example of a free-body force diagram. Figure 5 is not a free-body force diagram because it shows more than one object.

Figure 6 *Braking*

1 Describe and explain what happens to the glider in Figure 1 if the air track blower is switched off. [2 marks]

2 A jet plane lands on a runway and stops.
   a Give the direction of the resultant force on the plane as it lands. [1 mark]
   b Describe the resultant force on the plane when it has stopped. [1 mark]

3 A car is stuck in the mud. A tractor tries to pull it out.
   a The tractor pulls the car with a force of 250 N. Give the reason why the car does not move. [1 mark]
   b Increasing the tractor force to 300 N pulls the car steadily out of the mud at a constant speed and direction. Calculate the force of the mud on the car now. [1 mark]

4 a Copy the car in Figure 6 and show the weight of the car as a vector arrow midway between the wheels. [1 mark]
   b 🄷 Show the support forces of the road on each wheel as vector arrows. [1 mark]

Key points

- The resultant force is a single force that has the same effect as all the forces acting on an object.
- If the resultant force on an object is:
  - zero, the object stays at rest or at the same speed and direction
  - greater than zero, the speed or direction of the object will change.
- If two forces act on an object along the same line, the resultant force is:
  - their sum, if the forces act in the same direction
  - their difference, if the forces act in opposite directions.
- 🄷 A free-body force diagram of an object shows the forces acting on it.

P8.4 Centre of mass

Learning objectives

After this topic, you should know:

- what the centre of mass of an object is
- where the centre of mass of a metre ruler is
- about the centre of mass of an object suspended from a fixed point
- how to find the centre of mass of a symmetrical object.

The design of racing cars has changed a lot since the first models. Look at Figure 1, which shows examples of past and modern racing car designs. One thing that has not changed is the need to keep the car near to the ground. The weight of the car must be as low as possible. Otherwise, the car would overturn when going round corners at high speeds.

Figure 1 *Racing cars from the 1920s to the modern day*

You can think of the weight of an object as if it acts at a single point. This point is called the centre of mass (or the centre of gravity) of the object. The idea of centre of mass is very useful to designers and engineers. For example, the designer of a chair needs to use a force diagram showing its weight as well as all the other forces acting on it, to make sure the finished chair will not tip over when someone sitting on it leans back.

The centre of mass of an object is the point at which its mass can be thought of as being concentrated.

Suspended equilibrium

If you suspend an object and then release it, it will sooner or later come to rest with its centre of mass directly below the point of suspension, as shown in Figure 2a. The object is then in equilibrium, which means it is at rest. Its weight does not exert a turning effect on the object, because its centre of mass is directly below the point of suspension.

If the object is turned from this position and then released, it will swing back to its equilibrium position. This is because its weight has a turning effect that returns the object to equilibrium, as shown in Figure 2b. You say that the object is *freely suspended* if it returns to its equilibrium position after the turning force is taken away.

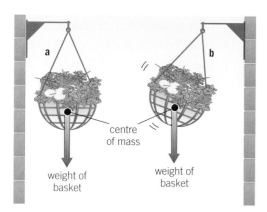

Figure 2 *Suspension* **a** *In equilibrium* **b** *Non-equilibrium*

The centre of mass of a symmetrical object

For a flat object that is symmetrical, its centre of mass is along the axis of symmetry. You can see this in Figure 3.

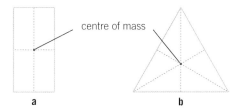

Figure 3 *Symmetrical objects*

If the object has more than one axis of symmetry, its centre of mass is where the axes of symmetry meet.

- A rectangle or a uniform ruler has two axes of symmetry, as shown Figure 3a. The centre of mass is where the axes meet.

- The equilateral triangle in Figure 3b has three axes of symmetry, each bisecting one of the angles of the triangle. The three axes meet at the same point. This is where the centre of mass of the triangle is.

Centre of mass

Figure 4 shows how to find the centre of mass of an irregular-shaped card.

1   Put a hole in one corner of the card and suspend the card from a rod.

2   Use a plumb line to draw a vertical line on the card from the rod.

3   Repeat the procedure, hanging the card from a different corner.

The point where the two lines meet is the centre of mass.

Use this method to find the centre of mass of a semicircular card of a radius 100 mm.

- Evaluate the accuracy of your experiment. For example, the card should balance on the flat end of a pencil placed directly under the card's centre of mass.

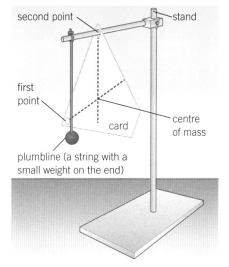

Figure 4 *Finding the centre of mass of a card*

1   Sketch each of the objects shown and mark its centre of mass.

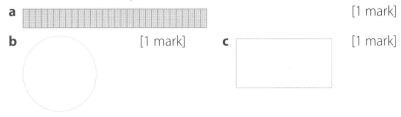

a   [1 mark]

b   [1 mark]   c   [1 mark]

2   Explain why a child sitting on a swing comes to rest directly below the top of the swing.   [2 marks]

3   Describe how you would find the centre of mass of a flat semicircular card.   [4 marks]

4   Look again at the flower baskets in Figure 2. Describe the resultant force on each basket. Give a reason for your answer in each case.   [2 marks]

Key points

- The centre of mass of an object is the point where its mass can be thought of as being concentrated.
- The centre of mass of a uniform ruler is at its midpoint.
- When an object is freely suspended, it comes to rest with its centre of mass directly underneath the point of suspension.
- The centre of mass of a symmetrical object is along the axis of symmetry.

P8.5 The parallelogram of forces

Learning objectives

After this topic, you should know:

- what the parallelogram of forces is
- what the parallelogram of forces is used for
- what is needed to draw a scale diagram of the parallelogram of forces
- how to use the parallelogram of forces to find the resultant of two forces.

Figure 1 *In tow*

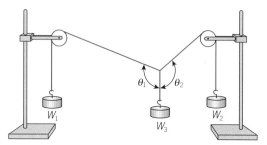

Figure 3 *The parallelogram of forces*

In Topic P8.2, you learnt how to find the resultant of two forces that act along the same line. What if the two forces do not act along the same line? Figure 2 shows a ship being towed by cables from two tugboats. The tension force in each cable pulls on the ship. The combined effect of these tension forces is to pull the vessel forwards. This is the resultant force.

Figure 2 shows how the two tension forces T_1 and T_2, represented as vectors, combine to produce the resultant force. The tension forces are drawn to scale as adjacent sides of a parallelogram. The angle between the two adjacent sides must be the same as the angle between the two forces. The resultant force is the diagonal of the parallelogram from the origin of T_1 and T_2. This geometrical method is called the **parallelogram of forces**.

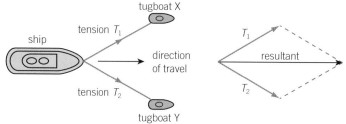

Figure 2 *Combining forces*

Investigating the parallelogram of forces

You can use weights and pulleys to demonstrate the parallelogram of forces (Figure 3). The tension in each string is equal to the weight it supports, either directly or over a pulley.

The point where the three strings meet is at rest. The string supporting the middle weight (W_3) is vertical. Using a protractor, you can measure angles θ_1 and θ_2 and note the values of the three known weights. You can then draw a scale diagram of a parallelogram so that:

- the line down the centre of the diagram represents the vertical line through the point where the three strings meet
- adjacent sides of the parallelogram at angles θ_1 and θ_2 to the vertical line represent the tensions in the strings supporting W_1 and W_2.

The resultant force of W_1 and W_2 represented by the diagonal line should be equal and opposite in direction to the vector representing W_3.

Make a model zip wire

Use a length of thin string and a weight hanger (or other suitable object) to make and test a model zip wire. Figure 4 shows the idea.

Release the weight hanger on the string at the top end and observe where it comes to rest.

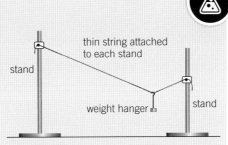

Figure 4 A model zip wire

Investigate how the height difference between the ends of the string affects the horizontal distance from the rest position of the hanger to one of the stands.

● Plot your results on a graph and discuss the effect of the height difference on the rest position of the hanger.

Safety: Make sure stands are clamped to the bench.

1 Figures 6 and 7 show examples where two forces act on an object X. In each case, work out the magnitude and direction of the resultant force on X.

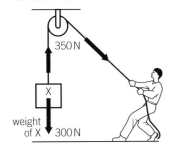

Figure 6

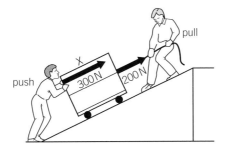

Figure 7

[2 marks]

2 A force of 3.0 N and a force of 4.0 N act on a point. Determine the magnitude and direction of the resultant of these two forces if the angle between their lines of action is:

a 90° [2 marks] **b** 60° [2 marks] **c** 45°. [2 marks]

3 In Figure 5, suppose the angle between the two sections of rope joined to the car had been 50° instead of 30°. Use the parallelogram of forces to find the maximum tension in the main tow rope. [2 marks]

4 In a model zip wire like the one shown in Figure 4, the two sections of the string are both at an angle of 20° to a horizontal line through the lowest point P of the string.

a Draw a diagram to show the line of action of the forces due to each string acting on point P. [1 mark]

b i The weight hanger has a weight of 2.0 N. Using a suitable scale, draw a vector arrow on your diagram to show the weight of the weight hanger. Label the scale on your diagram. [1 mark]

ii Use your diagram to find the tension in each section of the string. [3 marks]

A tow rope is attached to a car at two points 0.80 m apart. The two sections of rope joined to the car are the same length and are at 30° to each other (Figure 5). The pull on each attachment should not exceed 3000 N. Use the parallelogram of forces to determine the maximum tension in the main tow rope.

Solution

The maximum tension T in the main tow rope is the resultant force of the two 3000 N forces at 30° to each other.

Drawing the parallelogram of forces as shown in Figure 5 gives:

$$T = 5800 \text{ N}$$

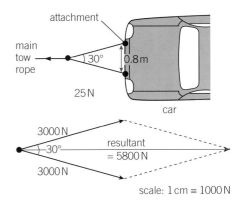

Figure 5 Using the parallelogram of forces

Key points

● The parallelogram of forces is a scale diagram of two force vectors.
● The parallelogram of forces is used to find the resultant of two forces that do not act along the same line.
● You will need a protractor, a ruler, a sharp pencil, and a blank sheet of paper.
● The resultant is the diagonal of the parallelogram that starts at the origin of the two forces.

P8.6 Resolution of forces

Learning objectives

After this topic, you should know:

- what is meant by resolution of a force
- how to resolve a force
- about the forces on an object in equilibrium
- how to use a force diagram to work out whether or not an object is in equilibrium.

Figure 1 *Cyclists on an uphill road*

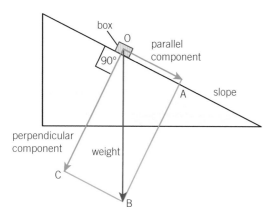

Figure 2 *Resolving a force*

Cyclists know that it is more difficult to travel uphill than it is to travel on a flat road (Figure 1). The reason is that the weight of the cyclist and the bicycle have a downhill effect. To understand this effect, consider a small box on a slope as shown in Figure 2. The weight of the box as a force vector is shown by the line labelled OB. You can think of the force vector as two parts or *components* – one force component acting down the slope, and the other force component acting perpendicular or normal to the slope. The process of looking at force in this way is called resolving a force into two components and is carried out as follows:

- A rectangle OABC is drawn on Figure 2. The weight of the box, OB, lies along the diagonal of the rectangle, and the rectangle's sides are parallel and perpendicular to the slope.

- Side OA of the rectangle lies along the slope, and represents the component of the weight acting down the slope. The box stays at rest on the slope and does not slide down, because friction acts on it. The component of weight acting down the slope is too small to overcome this frictional force. For the cyclist on an uphill road, the component of weight acting down the slope is the amount of force that the cyclist has to match in order to keep moving up the hill.

- Side OC of the rectangle OABC gives the component of the weight acting normal to the slope. This is the force pressing on the slope due to the box.

Test an incline

Use the arrangement in Figure 3 to measure the force F needed to keep a trolley in the same position on the inclined board.

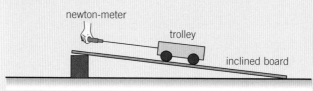

Figure 3 *Testing an incline*

1 Use a newton-meter to measure the weight W of the trolley and the force F.

Add weights to the trolley to find out how F changes as the weight of the trolley is increased.

2 Repeat the test with the board at a different angle to the laboratory bench.

Record all your measurements in a table, and plot a graph of your results.

- Write down the conclusions you make from your graph about the relationship between force F and weight W.

Safety: Beware of falling objects.

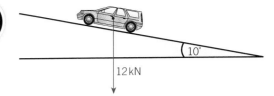

Figure 4 *Car parked on an uphill road*

Worked example

A car of weight 12 kN is parked on an uphill road. The road is inclined at an angle of 10° to the horizontal as shown in Figure 4.

a Use a geometrical method to find the component of the car's weight acting down the slope.

b Describe the force of friction of the road on the car tyres.

Solution

a See Figure 5. The ratio of the rectangle's small side to the diagonal = 1 : 6.0. Therefore, the parallel component of the weight = 12/6 kN = **2.0 kN**

b The frictional force = 2.0 kN acting up the slope.

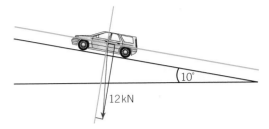

Figure 5

Equilibrium

An object at rest is in equilibrium. The key conditions for an object to be in equilibrium are:

● The resultant force on the object is zero

● The forces acting on the object have no overall turning effect.

To work out whether or not an object is in equilibrium:

● If the lines of action of the forces are parallel, the sum of the forces in one direction must be equal to the sum of the forces in the opposite direction. This means that the resultant force on the object is zero.

● If the lines of action of the forces are not all parallel, the forces can be resolved into two components along the same perpendicular lines. The components along each line must balance out if the resultant force is zero.

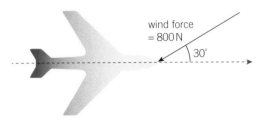

Figure 6 *Aircraft in level flight*

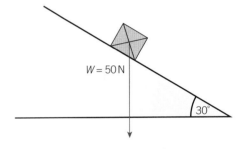

Figure 7 *Box resting on a slope*

1 An aircraft in level flight is travelling due east at a constant speed when it is acted on by a horizontal wind force of 800 N as shown in Figure 6. By resolving the wind force into two perpendicular components, determine the component of the wind force along the line in which the aircraft is moving. [2 marks]

2 A student pushed a trolley of weight 510 N up a slope that is inclined at 15° to the horizontal.
  a Determine the component of the trolley's weight down the slope. [2 marks]
  b The force exerted by the student was greater than the answer to part **a**. Give one possible reason for this difference. [1 mark]

3 Explain why a ladder placed against a wall would slide down the wall if the floor was too slippery. [2 marks]

4 A box of weight 50 N is at rest on a slope. The slope is at an angle of 30° to the horizontal as shown in Figure 7.
  a Copy the diagram and determine the components of the box's weight parallel and perpendicular to the slope. [3 marks]
  b Describe the friction between the box and the slope. [1 mark]

Key points

● Resolving a force means finding perpendicular components that have a resultant force that is equal to the force.

● To resolve a force in two perpendicular directions, draw a rectangle with adjacent sides along the two directions so that the diagonal represents the force vector.

● For an object in equilibrium, the resultant force is zero.

● An object at rest is in equilibrium because the resultant force on it is zero.

P8 Forces in balance

Summary questions

1 Figure 1 shows an iron bar suspended at rest from a spring balance that reads 1.6 N.

Figure 1

a i Calculate the magnitude and the direction of the force on the spring balance due to the iron bar. [1 mark]

ii Calculate the weight of the bar in newtons. [1 mark]

b When a magnet is held under the iron bar, the spring balance reading increases to 2.0 N. Calculate the magnitude and the direction of:

i the force on the iron bar due to the magnet. [1 mark]

ii the force on the magnet due to the iron bar. [1 mark]

2 A suitcase of weight 180 N lies on a level floor.

a i What is the magnitude and direction of the force of the suitcase on the floor? [1 mark]

ii What is the magnitude and direction of the force of the floor on the suitcase? [1 mark]

b A child of weight 40 N sits on the suitcase.

i What is the magnitude and direction of the force of the suitcase on **i** the child [1 mark]

ii the floor? [1 mark]

3 Figure 2 shows a stationary helium-filled balloon attached to a vertical thread. Because helium is lighter than air, an upward force, or upthrust, acts on the balloon. The lower end of the string is attached to a weight on a table.

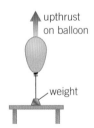

Figure 2

a i What can you say about the resultant force on the balloon? [1 mark]

ii Which force is greater: the gravitational force on the balloon or the upthrust? Give a reason for your answer. [2 marks]

b The centre of mass of the balloon is approximately midway between the top and bottom of the balloon.

i Draw the balloon and mark the centre of mass **C** of the balloon on your sketch. [1 mark]

ii Use your sketch to draw a free-body diagram of the forces acting on the balloon. [2 marks]

c Describe and explain what would happen to the balloon if the thread was cut. [2 marks]

4 A yacht is pushed along at constant velocity by the wind.

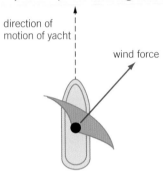

Figure 3

a What can you say about the resultant force on yacht? [1 mark]

b The wind force is 180 N and its direction is at an angle of 25° to the direction of motion of the yacht. By resolving the wind force parallel and perpendicular to the direction of motion, determine the component of the wind force in the direction of motion. [3 marks]

5 a **H** Look at Figure 4. Two tugboats are used to pull a ship. Each tugboat exerts a force of 7200 N on the ship at an angle of 45° between their cables, as shown in the figure. Use the parallelogram of forces to find the magnitude of the resultant of the tugboat forces on the ship. [2 marks]

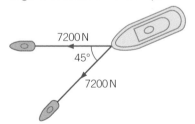

Figure 4

b The ship moves at a constant speed and direction because it is acted on by a drag force. Explain why the drag force has this effect on the ship. [2 marks]

Practice questions

01.1 Forces can be either contact forces or non-contact forces. Give the name of one contact force and one non-contact force. [2 marks]

01.2 **Figure 1** shows a water skier being pulled by a speed boat.

Figure 1

The motive force of the speed boat is 20 000 N. Copy and complete the sentence using the correct words from the box.

| less than | equal to | greater than |

When the water skier accelerates through the water, the resistive force of the water is _____ 20 000 N. [1 mark]

01.3 Describe what happens to the speed and resistive force on the water skier as she accelerates through the water. [3 marks]

02 **(H)** A drone is a radio-controlled flying device.

Figure 2

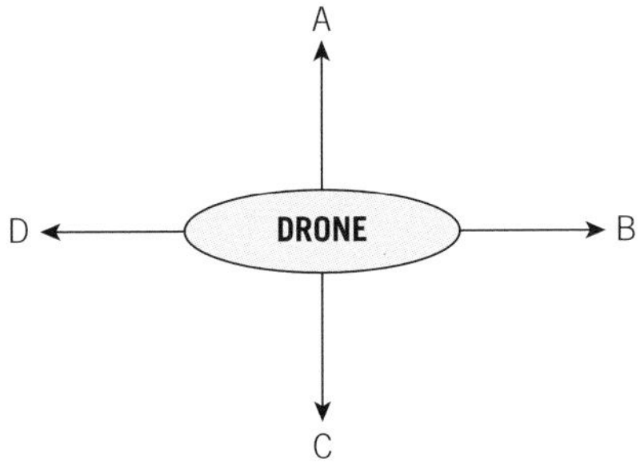

A
↑
D ◄─── DRONE ───► B
↓
C

02.1 Copy and complete the sentences using correct answers from the box.

| force A | force B | force C | force D |

When the drone is flying at a constant height _____ and _____ are equal and opposite. When the drone is flying at constant speed _____ and _____ are equal and opposite. [2 marks]

02.2 A video camera is attached to the drone. The drone is used to film alligators in swampland. Give two advantages of using a drone for this purpose. [2 marks]

02.3 Some people object to the use of drones in public places. Suggest two reasons why the use of a drone may be a problem. [2 marks]

03 A space craft has boosters on each of its sides. When lining up to dock with a space station, two boosters fire providing forces of 500 N and 750 N. The forces act at a right angle to each other.

03.1 Draw a free body diagram showing the forces acting on the space craft. You can assume the only forces are those provided by the boosters. [2 marks]

03.2 Draw a vector diagram to determine the magnitude and direction of the resultant force on the space craft.
Use graph paper. [4 marks]

Learning objectives

After this topic, you should know:

- how speed is calculated for an object moving at constant speed
- how to use a distance–time graph to determine whether an object is stationary or moving at constant speed
- what the gradient of the line on a distance–time graph can tell you
- how to use the equation for constant speed to calculate distance moved or time taken.

Figure 1 *The Budweiser Rocket attempting the land speed record in 1979. Thust SSC driven by Andy Green in 1997 achieved a speed of 341 m/s, and still held the world land speed record in 2015*

Rearranging the speed equation

The equation for speed can be written as:

$$v = \frac{s}{t}$$

where *v* is the speed, *s* is the distance travelled, and *t* is the time taken. If you know two of these three quantities, you can find the third by using $v = \frac{s}{t}$ or by rearranging it to give:

$$s = v\,t \quad \text{or} \quad t = \frac{s}{v}.$$

Next time you are travelling on a motorway, look for the marker posts positioned every kilometre. If you are a passenger in a car on a motorway, you can use these posts to check the speed of the car. You need to time the car as it passes each post. Table 1 shows some measurements made on a car journey.

Table 1 *Measurements made on a car journey*

Distance in metres	0	1000	2000	3000	4000	5000	6000
Time in seconds	0	40	80	120	160	200	240

The measurements in Table 1 are plotted on a graph of distance against time in Figure 2.

- the car took 40 s to go from each marker post to the next. So its speed was constant (or uniform).
- the car went a distance of 25 metres every second (= 1000 metres ÷ 40 seconds). So its speed was 25 metres per second.

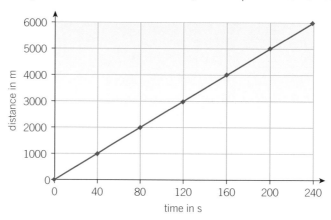

Figure 2 *A distance–time graph*

If the car had travelled faster, it would have gone further than 1000 metres every 40 seconds. So the line on the graph would have been steeper. In other words, the **gradient** of the line would have been greater.

The gradient of a line on a distance–time graph represents speed.

Equation for constant speed

For an object moving at constant speed, you can calculate its speed using the equation:

speed, *v* (metres per second, m/s) = $\dfrac{\text{distance travelled, } s \text{ (metres, m)}}{\text{time taken, } t \text{ (seconds, s)}}$

The scientific unit of speed is the metre per second, usually written as metre/second or m/s.

This equation can also be used to calculate the average speed of an object whose speed varies. For example, if a motorist in a traffic queue took 50 s to travel a distance of 300 m, the car's average speed was 6.0 m/s (= 300 m ÷ 50 s).

Speed in action

Long-distance vehicles are fitted with recorders called tachographs. These can check that their drivers do not drive for too long. Look at the distance–time graph in Figure 3 for three lorries, X, Y, and Z, on the same motorway.

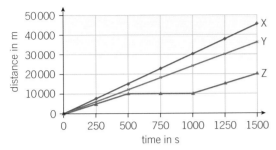

Figure 3 *Comparing distance–time graphs*

- X went fastest because it travelled further than Y or Z in the same time.

- Y travelled more slowly than X. You can see this from the graph because the line for Y has a smaller gradient than the line for X. Also you can see that Y travelled 30 000 metres in 1250 seconds. So its speed was:

    distance ÷ time = 30 000 m ÷ 1250 s = 24 m/s.

- Z travelled the least distance. It stopped for some of the time. This is shown by the flat section of the graph (from 500 s to 1000 s). The gradient is zero in this section because Z was stationary. Its speed was zero during this time. When it was moving, its speed was also less than that of X or Y. You can see this from the graph because the gradient of the line for Z when it was moving is less than the gradient of the lines for X and Y.

1 a For an object travelling at constant speed:
  i describe the distance it travels every second. [1 mark]
  ii describe the gradient of its distance–time graph. [1 mark]
  b Look at the distance–time graphs in Figure 3.
  i Calculate the speed of X. [2 marks]
  ii Calculate how long Z stopped for. [1 mark]
  iii Calculate the *average* speed of Z for the 1500 s journey. [2 marks]

2 A vehicle on a motorway travels 1800 m in 60 s. Calculate:
  a the average speed of the vehicle in m/s. [2 marks]
  b how far it would travel if it travelled at this speed for 300 s. [2 marks]
  c how long it would take to travel a distance of 3300 m at this speed. [2 marks]

3 A car on a motorway travels a certain distance in six minutes at a speed of 21 m/s. A coach takes seven minutes to travel the same distance. Calculate the distance and the speed of the coach. [2 marks]

4 A train takes 2 hours and 40 minutes to travel a distance of 360 km.
  a Calculate the average speed of the train in metres per second on this journey. [2 marks]
  b The train travelled at a constant speed of 40 m/s for a distance of 180 km. Calculate the time taken in minutes for this section of the journey. [2 marks]

Synoptic links

For more information on rearranging equations, see Maths skills for Physics.

Study tip

If time is given in minutes or hours in these calculations, always convert it into seconds.

Key points

- The speed of an object is: $v = \frac{s}{t}$.
- The distance–time graph for any object that is:
    - stationary, is a horizontal line
    - moving at constant speed, is a straight line that slopes upwards.
- The gradient of a distance–time graph for an object represents the object's speed.
- The speed equation $v = \frac{s}{t}$ can be rearranged to give:
    $$s = v\,t \text{ or } t = \frac{s}{v}$$

P9.2 Velocity and acceleration

Learning objectives

After this topic, you should know:

- the difference between speed and velocity
- how to calculate the acceleration of an object
- the difference between acceleration and deceleration.

Figure 1 *You experience plenty of changes in velocity on a corkscrew ride*

Synoptic link

Velocity and displacement are vector quantities because they have magnitude (size) and direction. Speed is a scalar quantity because it has magnitude only. See Topic P8.1.

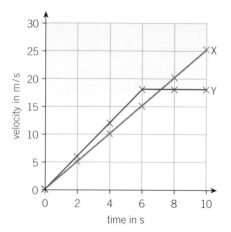

Figure 2 *Velocity–time graph for cars X and Y*

In fairground rides that throw you round and round, your speed and direction of motion keep changing. The word **velocity** is used for speed in a given direction.

Velocity is speed in a given direction.

- Two moving objects can have the same speed but different velocities. For example, a car travelling north at 30 m/s on a motorway has the same speed as a car travelling south at 30 m/s. But their velocities are not the same, because they are moving in opposite directions.

- An object moving round in a circle has a direction of motion that changes continuously as it goes round. So its velocity is not constant even if its speed is constant. For example, a car travelling round a roundabout at constant speed has a continually changing velocity. This is because the direction in which it is moving is continually changing.

An object that travels at constant velocity travels at a constant speed without changing its direction. So it travels in a straight line in a given direction. The word **displacement** is used for the distance travelled in a given direction.

Acceleration

A car maker claims that their new car accelerates more quickly than any other new car. A rival car maker is not pleased by this claim and issues a challenge. Each car in turn is tested on a straight track with a velocity recorder fitted. The results are shown in Table 1.

Table 1 *Results of the car velocity tests*

Time from a standing start in seconds (s)	0	2	4	6	8	10
Velocity of car X in metres per second (m/s)	0	5	10	15	20	25
Velocity of car Y in metres per second (m/s)	0	6	12	18	18	18

Which car has a greater **acceleration**? The results are plotted on the velocity–time graph in Figure 2. You can see that the velocity of Y goes up from zero faster than the velocity of X does. So Y has a greater acceleration in the first six seconds. After that, the velocity of Y is constant so its acceleration is zero after six seconds.

The acceleration of an object is its change of velocity per second.
The unit of acceleration is the metre per second squared, or m/s².

Any object with a changing velocity is accelerating. You can work out its average acceleration a for a change of velocity Δv in time t using the equation:

acceleration, a (metres per second squared, m/s²)

$$= \frac{\textbf{change in velocity, } \Delta v \text{ (metres per second, m/s)}}{\textbf{time taken for the change, } t \text{ (seconds, s)}}$$

Higher

For an object that accelerates steadily from an initial velocity u to a final velocity v, in time t:

its change of velocity Δv = final velocity v − initial velocity u.

So you can write the equation for acceleration as:

$$a = \frac{v - u}{t}$$

Deceleration

A car decelerates when the driver brakes. The term **deceleration** or negative acceleration is used for any situation where an object slows down.

Worked example

A car moving at a velocity of 28 m/s brakes and stops in 8.0 s.

Calculate its deceleration.

Solution

initial velocity u = 28 m/s, final velocity v = 0, time taken t = 8.0 s

$$\text{acceleration } a = \frac{\text{change in velocity}}{\text{time taken}} = \frac{v - u}{t}$$
$$= \frac{0\,\text{m/s} - 28\,\text{m/s}}{8.0\,\text{s}} = -3.5\,\text{m/s}^2$$

The deceleration is therefore 3.5 m/s².

Worked example

In Figure 2, the velocity of Y increases from 0 to 18 m/s in 6.0 s.

Calculate its acceleration.

Solution

Change of velocity = $v - u$
$$= 18\,\text{m/s} - 0\,\text{m/s}$$
$$= 18\,\text{m/s}$$

Time taken t = 6.0 s

$$\text{Acceleration } a = \frac{\text{change in velocity}}{\text{time taken}}$$
$$= \frac{v - u}{t}$$
$$= \frac{18\,\text{m/s}}{6.0\,\text{s}}$$
$$= 3.0\,\text{m/s}^2$$

Study tip

Be careful with units, especially the unit of acceleration. The unit is m/s², that is, the change in speed measured in m/s that occurs every second.

Study tip

A minus value for acceleration means that the object is slowing down.

If you are talking about deceleration, you do not need to include a minus sign.

1 **a** Compare speed and velocity. [1 mark]
 b A car on a motorway is travelling at a constant speed of 30 m/s when it overtakes a lorry travelling at a speed of 22 m/s. If both vehicles maintain their speeds, calculate how far ahead of the lorry the car will be after 300 s. [2 marks]

2 The velocity of a car increased from 8 m/s to 28 m/s in 16 s without change of direction. Calculate its acceleration. [2 marks]

3 The driver of a car increased the speed of the car as it joined the motorway. It then travelled at constant velocity before slowing down as it left the motorway at the next junction.
 a i Explain when the car decelerated. [1 mark]
  ii Explain when the acceleration of the car was zero. [1 mark]
 b When the car joined the motorway, it accelerated from a speed of 7.0 m/s for 10 s at an acceleration of 2.0 m/s². Calculate its speed at the end of this time. [3 marks]

4 A sprinter in a 100 m race accelerated from rest and reached a speed of 9.2 m/s in the first 3.1 s.
 a Calculate the acceleration of the sprinter in this time. [2 marks]
 b The sprinter continued to accelerate to top speed and completed the race in 10.4 s. Calculate the sprinter's average speed. [2 marks]

Key points

- Velocity is speed in a given direction.
- A vector is a physical quantity that has a direction as well as a magnitude. A scalar is a physical quantity that has a magnitude only and does not have a direction.
- The acceleration of an object is $a = \frac{\Delta v}{t}$
- Deceleration is the change of velocity per second when an object slows down.

P9.3 More about velocity–time graphs

Learning objectives

After this topic, you should know:

- how to measure velocity changes
- what a horizontal line on a velocity–time graph tells you
- how to use a velocity–time graph to work out whether an object is accelerating or decelerating
- **H** what the area under a velocity–time graph tells you.

Investigating acceleration

Use a motion sensor and a computer to find out how the steepness of a runway affects a trolley's acceleration.

- In this investigation, name:
  - **i** the independent variable
  - **ii** the dependent variable.
- Describe the relationship you find between the two variables.

Safety: Use foam or an empty cardboard box to stop the trolley falling off the bench. Mind your feet!

Investigating acceleration

You can use a motion sensor linked to a computer to record how the velocity of an object changes. Figure 1 shows how you can do this using a trolley as the moving object. The computer can also be used to display the measurements as a velocity–time graph.

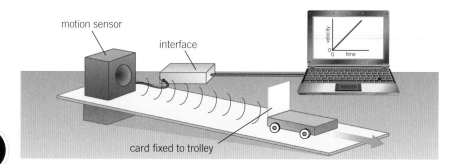

Figure 1 *Investigating velocity–time graphs using a computer*

Test A: If you let the trolley accelerate down the runway, its velocity increases with time. Look at the velocity–time graph from a test run on the laptop screen in Figure 1.

- The line goes up because the velocity increases with time. So it shows that the trolley was accelerating as it ran down the runway.
- The line is straight, which tells you that the increase in velocity was the same every second. In other words, the acceleration of the trolley was constant.

Test B: If you make the runway steeper, the trolley accelerates faster. This would make the line on the graph in Figure 1 steeper than for test A. So the acceleration in test B is greater.

These tests show that:

> **the gradient of the line on a velocity–time graph represents acceleration.**

Braking

Braking reduces the velocity of a vehicle. Look at the graph in Figure 2. It is the velocity–time graph for a vehicle that brakes and stops at a set of traffic lights. The velocity is constant until the driver applies the brakes.

- The section of the graph with the horizontal line shows constant velocity. The gradient of the line is zero, so the acceleration in this section is zero.
- When the brakes are applied, the vehicle decelerates, and its velocity decreases to zero. The gradient of the line is negative in this section. So the acceleration is negative.

The area under the line on a velocity–time graph represents distance travelled in a given direction (or displacement).

Look at Figure 2 again.

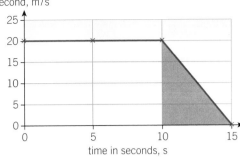

Figure 2 *Braking*

- Before the brakes are applied, the vehicle moves at a velocity of 20 m/s for 10 s. So it travels 200 m in this time (= 20 m/s × 10 s). This distance is represented on the graph by the area under the line from 0 s to 10 s. This is the rectangle shaded yellow on the graph.

- When the vehicle decelerates in Figure 2, its velocity drops from 20 m/s to 0 m/s in 5 s. You can work out the distance travelled in this time from the area of the purple triangle in Figure 2. This area is:

$\frac{1}{2}$ × the height of the triangle × the base of the triangle

$= \frac{1}{2} \times 20\,m/s \times 5\,s = 50\,m.$

So the vehicle must have travelled a distance of 50 m when it was decelerating.

Speed and velocity can also be described in kilometres per hour (km/h).

Because 1000 m = 1 km, and 3600 s = 1 hour, then a speed of 1 km/h is equal to 1000 m ÷ 3600 s = 0.278 m/s.

If speed and velocity are given in kilometres per hour (km/h), their values must be converted to metres per second before plotting them on a graph or using them in an equation.

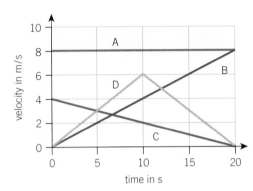

Figure 3

If you are drawing a straight line graph, always use a ruler.

1 Match each of the following descriptions (**i** to **iv**) to one of the lines, labelled **A**, **B**, **C**, and **D** on the velocity–time graph (Figure 3).
   i Accelerated motion throughout.
   ii Zero acceleration.
   iii Accelerated motion, then decelerated motion.
   iv Deceleration throughout. [2 marks]

2 Ⓗ Look at Figure 3.
   a Identify the line representing the object that travelled:
      i the furthest distance [1 mark]
      ii the least distance. [1 mark]
   b Which object, **B** or **D**, travelled further? [1 mark]

3 Ⓗ Look again at Figure 3. Show that the object that produced the data for line A (the horizontal line) travelled a distance of 160 m. [1 mark]
   b Determine the distance travelled by object B. [1 mark]

4 Ⓗ Look again at the graph in Figure 3.
   a Calculate the distance travelled by object **C**. [2 marks]
   b Calculate the difference in the distances travelled by **A** and **D**. [3 marks]

Key points

- A motion sensor linked to a computer can be used to measure velocity changes.
- The gradient of the line on a velocity–time graph represents acceleration.
- If a velocity–time graph is a horizontal line, the acceleration is zero.
- A positive gradient on a velocity–time graph represents positive acceleration, a negative gradient represents deceleration.
- Ⓗ The area under the line on a velocity–time graph represents distance travelled.

P9.4 Analysing motion graphs

Learning objectives

After this topic, you should know:

- how to calculate acceleration from a velocity–time graph
- **H** how to calculate distance from a velocity–time graph.
- **H** how to calculate speed from a distance–time graph:
  - where the speed is constant
  - where the speed is changing.

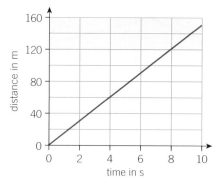

Figure 1 *A distance–time graph for constant speed*

Acceleration

Use the graph in Figure 3 to find the acceleration of the object.

Solution

The height of the triangle represents an increase of velocity of 8 m/s (= 12 m/s – 4 m/s).

The base of the triangle represents a time of 10 s.

So the acceleration = $\dfrac{\text{change of velocity}}{\text{time taken}}$

$= \dfrac{8\,\text{m/s}}{10\text{s}} = 0.8\,\text{m/s}^2$

Using distance–time graphs

For an object moving at constant speed, the distance–time graph is a straight line sloping upwards (Figure 1).

The speed of the object is represented by the gradient of the line. To find the gradient, you need to draw a triangle under the line, as shown in Figure 2. The height of the triangle represents the distance travelled, and the base of the triangle represents the time taken. So:

$$\text{the gradient of the line} = \frac{\text{the height of the triangle}}{\text{the base of the triangle}}$$

and this represents the object's speed.

For a moving object with changing speed, the distance–time graph is not a straight line. The red line in Figure 2 shows an example.

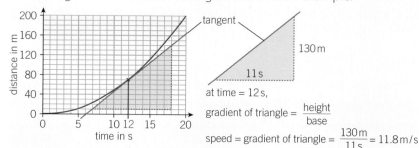

at time = 12 s,

gradient of triangle = $\dfrac{\text{height}}{\text{base}}$

speed = gradient of triangle = $\dfrac{130\,\text{m}}{11\,\text{s}}$ = 11.8 m/s

Figure 2 *A distance–time graph for changing speed*

In Figure 2, the gradient of the line increases, so the object's speed must have increased. You can find the speed at any point on the line by drawing a tangent to the line at that point, as shown in Figure 2. The **tangent** to the curve is a straight line that touches the curve without cutting through it (in this case at 12 s). The gradient of the tangent is equal to the speed of the object at that instant in time.

Using velocity–time graphs

Figure 3 shows the velocity–time graph of an object moving with a constant acceleration. Its velocity increases at a steady rate. So the graph shows a straight line that has a constant gradient.

To find the acceleration from the graph, remember that the gradient of the line on a velocity–time graph represents the acceleration.

In Figure 3, the gradient is given by the height divided by the base of the triangle under the line. The height of the triangle represents the change of velocity, and the base of the triangle represents the time taken.

So the gradient represents the acceleration, because:

$$\text{acceleration} = \frac{\text{change of velocity}}{\text{time taken}}$$

Higher

To find the distance travelled from the graph, remember that the area under a line on a velocity–time graph represents the distance travelled. The shape under the line in Figure 3 is a triangle on top of a rectangle. So the distance travelled is represented by the area of the triangle plus the area of the rectangle under it. Prove for yourself that the triangle represents a distance travelled of 40 m and that the rectangle also represents a distance of 40 m. So the total distance travelled is 80 m (= 40 m + 40 m).

Distance, velocity, and acceleration

In Figure 3, the total distance travelled is 80 m, and the time taken t is 10 s. So the average velocity is 8 m/s, which is equal to $\frac{1}{2}(u + v)$, where the initial velocity $u = 4$ m/s, and the final velocity $v = 12$ m/s. So the distance travelled $s = \frac{1}{2}(u + v) \times t$. Because the acceleration $a = \frac{v - u}{t}$, then

$$a \times s = \frac{v - u}{t} \times \frac{1}{2}(u + v) \times t = \frac{1}{2}(v^2 - u^2)$$

Rearranging this equation gives $v^2 - u^2 = 2as$
This equation is useful for calculations where the time taken is not given and the acceleration is constant. You *do not* need to know how to prove this equation.

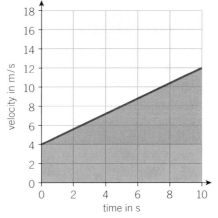

Figure 3 *A velocity–time graph for constant acceleration*

Study tip

Make sure that you know whether you are dealing with a distance–time graph or a velocity–time graph. The gradients of these two types of graph represent different quantities.

1 **a** Determine the speed of the object shown on the graph in Figure 1. [2 marks]
  b Describe how the speed of the object shown in Figure 2 changes. [2 marks]

2 The graph in Figure 4 shows how the velocity of a cyclist on a straight road changes with time.
  a Describe the motion of the cyclist. [2 marks]
  b Use the graph in Figure 4 to determine the acceleration of the cyclist and the distance travelled in:
   i the first 40 s [4 marks] **ii** the following 20 s. [4 marks]
  c 🅗 Calculate the average speed of the cyclist over the journey. [2 marks]

3 In a motorcycle test, the speed from rest was recorded at intervals in Table 1.

Table 1 *Motorcycle test results*

Time in s	0	5	10	15	20	25	30
Velocity in m/s	0	10	20	30	40	40	40

  a Plot a velocity–time graph of these results. [3 marks]
  b Calculate the initial acceleration of the motorcycle. [2 marks]
  c 🅗 Calculate how far the motorcycle moved in:
   i the first 20 s **ii** the following 10 s. [4 marks]
4 Use Table 1 and the equation $v^2 = u^2 + 2as$ to calculate the velocity of the motorcycle after 1.0 km from the start if it had kept the same acceleration as it had during the first 20 s. [2 marks]

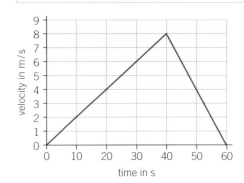

Figure 4 *See Summary question 2*

Key points

- The speed of an object moving at constant speed is given by the gradient of the line on its distance–time graph.
- The acceleration of an object is given by the gradient of the line on its velocity–time graph.
- 🅗 The distance travelled by an object is given by the area under the line on its velocity–time graph.
- 🅗 The speed, at any instant in time, of an object moving at changing speed is given by the gradient of the tangent to the line on its distance–time graph.

P9 Motion

Summary questions

1 A model car travels round a circular track at constant speed.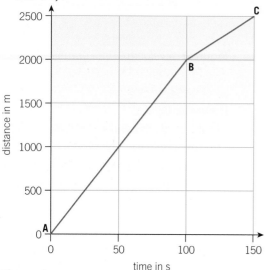
   If you were given a stopwatch, a marker, and a tape measure, explain how you would measure the speed of the car. [5 marks]

2 Figure 1 shows the distance–time graph for a car on a motorway.

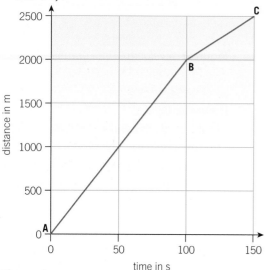

Figure 1

a Determine which part of the journey was faster – **A** to **B**, or **B** to **C**. [1 mark]

b i Calculate the speed of the car between **A** and **B**. [2 marks]

   ii Calculate the speed of the car between **B** and **C**. [2 marks]

c If the car had travelled the whole distance of 2500 m at the same speed as it travelled between **A** and **B**, calculate how long the journey would have taken. [2 marks]

3 Figure 2 shows a distance–time graph for a motorcycle approaching a speed limit sign.

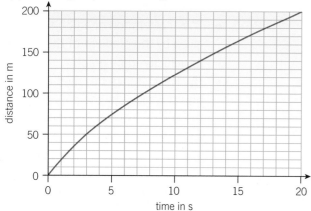

Figure 2

a Describe how the speed of the motorcycle changed with time. [2 marks]

b **H** Use the graph to determine the speed of the motorcycle:
   i initially   ii 10 s later. [5 marks]

4 a A car took 10 s to increase its velocity from 5 m/s to 30 m/s. Calculate its acceleration. [2 marks]

b The graph (Figure 3) shows how the velocity of the car changed with time during the 10 s.

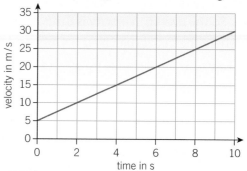

Figure 3

i **H** Calculate how far the car travelled in this time. [3 marks]

ii Calculate the average speed of the car in this time. [2 marks]

5 Table 1 shows how the velocity of a train changed as it travelled from one station to the next.

Table 1

Time in s	0	20	40	60	80	100	120	140	160
Velocity in m/s	0	5	10	15	20	20	20	10	0

a Plot a velocity–time graph using this data. [3 marks]

b Calculate the train's acceleration in each of the three parts of the journey. [5 marks]

c **H** Calculate the total distance travelled by the train. [4 marks]

d Show that the average speed for the train's journey was 12.5 m/s. [2 marks]

6 A water skier started from rest and accelerated steadily to 12 m/s in 15 s, then travelled at constant speed for 45 s, before slowing down steadily and coming to a halt 90 s after she started.

a Draw a velocity–time graph for this journey. [3 marks]

b Calculate the acceleration of the water skier in the first 15 s. [2 marks]

c Calculate the deceleration of the water skier in the final 30 s. [2 marks]

d **H** Calculate the total distance travelled by the water skier. [4 marks]

Practice questions

01   An electric truck is used to deliver parcels to
different locations in a warehouse.
The distance–time graph shows the journey taken
by the truck.

Figure 1

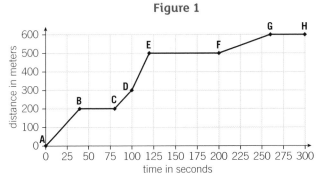

01.1   Determine the total distance travelled by the truck.
[1 mark]

01.2   Determine the total time taken to offload all the
parcels.   [2 marks]

01.3   Between which two points was the truck travelling
at the greatest speed?
Give a reason for your answer.   [2 marks]

01.4   Between which two points did the truck change
speed without stopping?   [2 marks]

01.5   Calculate the average speed of the truck over the
complete journey.
Write down the equation you use.   [3 marks]

01.6   Name one advantage of using an electric truck
rather than a diesel truck in a warehouse.   [1 mark]

02.1   Describe the difference between the terms speed
and velocity.   [2 marks]
A firework rocket is sent up into the air. The
velocity–time graph of the journey is shown.

Figure 2

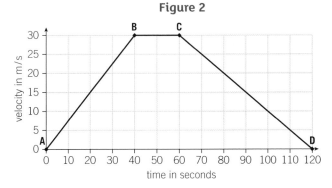

02.2   Calculate the average acceleration of the rocket
between 0 s and 40 s.
Give the unit.   [3 marks]

02.3   🅗   Use the graph in Figure 2 to calculate the total
distance the rocket travels.   [2 marks]

02.4   Copy and complete the sentences using the correct
words from the box.

> stationary   decelerating   accelerating
> travelling at a constant speed

The rocket is _____ between
points A and B.
The rocket is _____ between
points B and C.
The rocket is _____ between
points C and D.   [3 marks]

03   Engineers are testing a new water slide in an
aqua park. They are using a plastic dummy and
measuring the time it takes for the dummy to travel
from top to bottom. The rate of running water
through the slide is being changed.

03.1   Name the independent variable.   [1 mark]

03.2   Name the dependent variable.   [1 mark]

03.3   Suggest one reason why this is not a suitable test.
[1 mark]

03.4   In one test the dummy is sliding at 10.0 m/s at the
end of the slide. The total slide length is 20 m.
Calculate the acceleration of the dummy through
the slide. Use the equation:
$v^2 - u^2 = 2as$ [2 marks]

04   Some physical quantities are called scalars and
some physical quantities are called vectors.
Copy **Table 1**. Tick **one** box in each line.

Table 1

Physical Quantity	Scalar	Vector
speed		
mass		
acceleration		
weight		
energy		

[5 marks]

05   Competitors are taking part in a triathlon race
which involves running, swimming and cycling in
immediate succession over a 50 km distance.
Describe the factors that may affect the eventual
outcome of the race.
Your answer should include typical speed values at
various stages and what other factors may affect
the time of all the competitors in the race.   [4 marks]

Learning objectives

After this topic, you should know:

- how the acceleration of an object depends on the size of the resultant force acting upon it
- the effect that the mass of an object has on its acceleration
- how to calculate the resultant force on an object from its acceleration and its mass
- **H** what is meant by the inertia of an object.

Synoptic links

Calculating acceleration from a velocity–time graph was covered in Topic P9.4.

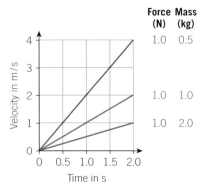

Force (N)	Mass (kg)
1.0	0.5
1.0	1.0
1.0	2.0

Figure 2 *Velocity–time graph for different combinations of force and mass*

Rearranging resultant force equation

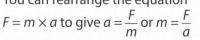

You can rearrange the equation $F = m \times a$ to give $a = \dfrac{F}{m}$ or $m = \dfrac{F}{a}$

You can use the apparatus shown in Figure 1 to accelerate a trolley with a constant force.

Investigating force and acceleration

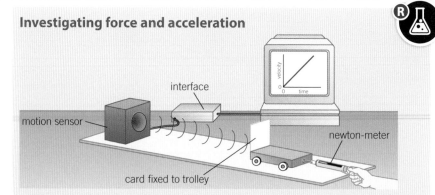

Figure 1 *Investigating the link between force and motion*

Use the newton-meter to pull the trolley along a known distance with a constant force.

A motion sensor and a computer can be used to record the velocity of the trolley as it accelerates.

- Predict what will happen to the acceleration of the trolley if the force is increased or decreased.

You can double or treble the total moving mass of the trolley by using double-deck and triple-deck trolleys.

- Predict what will happen to the acceleration of the trolley if the mass of the trolley is increased.

Safety: Protect your bench and feet from falling trolleys.

You can display the results of this investigation as a velocity–time graph on the computer screen. Figure 2 shows velocity–time graphs for different trolley masses. You can work out the acceleration from the gradient of the line. Some typical results using different forces and masses are given in Table 1.

Table 1 *Typical results for investigating force and acceleration*

Resultant force in N	0.5	1.0	1.5	2.0	4.0	6.0
Mass in kg	1.0	1.0	1.0	2.0	2.0	2.0
Acceleration in m/s²	0.5	1.0	1.5	1.0	2.0	3.0
Mass × acceleration in kg m/s²	0.5	1.0	1.5	2.0	4.0	6.0

The results show that the resultant force, the mass, and the acceleration are linked by the equation:

resultant force, F	=	mass, m	×	acceleration, a
(newtons, N)		(kilograms, kg)		(metres per second squared, m/s²)

Newton's second law

Newton's second law of motion says that the acceleration of an object is:

- proportional to the resultant force on the object
- inversely proportional to the mass of the object.

So the resultant force is proportional to the object's mass multiplied by its acceleration. You can write this as $F \propto ma$, where the symbol $\propto$ means 'is proportional to'. You can see from the equation for resultant force that 1 N is the force that gives a 1 kg mass an acceleration of $1 \, m/s^2$.

Inertia

Higher

A resultant force is needed to change the velocity of an object. The tendency of an object to stay at rest or to continue in uniform motion (i.e., moving at constant velocity) is called its **inertia**. The inertial mass of an object is a measure of the difficulty of changing the object's velocity.

Inertial mass can be defined as $\dfrac{\text{force}}{\text{acceleration}}$.

Speeding up or slowing down

If the velocity of an object changes, it must be acted on by a resultant force. Its acceleration is always in the same direction as the resultant force.

- The velocity of the object increases (i.e., it accelerates) if the resultant force is in the *same* direction as the velocity. You say that its acceleration is positive because it is in the same direction as its velocity.

- The velocity of the object decreases (i.e., it decelerates) if the resultant force is in the *opposite* direction to its velocity. You say that its acceleration is negative because it is in the opposite direction to its velocity.

1 a Calculate the resultant force on a sprinter of mass 80 kg who accelerates at $8 \, m/s^2$.   [1 mark]
   b Calculate the acceleration of a car of mass 800 kg acted on by a resultant force of 3200 N.   [1 mark]

2 Copy and complete the table.   [5 marks]

	Force in N	Mass in kg	Acceleration in m/s²
a		20	0.80
b	200		5.0
c	840	70	
d		0.40	6.0
e	5000		0.20

3 A car and a trailer have a total mass of 1500 kg.
   a Calculate the force needed to accelerate the car and the trailer at $2.0 \, m/s^2$.   [1 mark]
   b The mass of the trailer is 300 kg. Determine:
      i   the force of the tow bar on the trailer   [1 mark]
      ii the resultant force on the car.   [2 marks]

4 A constant force was used as in Figure 1 to accelerate a trolley from rest. The acceleration of the trolley was $0.60 \, m/s^2$. A mass of 0.5 kg was then fixed onto the trolley, and the same force as before gave it an acceleration of $0.48 \, m/s^2$.
   a Explain why the acceleration in the second case was less than before.   [2 marks]
   b Use the data above to calculate the mass of the trolley.   [2 marks]

Worked example

Calculate the resultant force on an object of mass 6.0 kg when it has an acceleration of $3.0 \, m/s^2$.

Solution

resultant force = mass × acceleration
$$= 6.0 \, kg \times 3.0 \, m/s^2$$
$$= 18.0 \, N$$

Worked example

Calculate the acceleration of an object of mass 5.0 kg acted on by a resultant force of 40 N.

Solution

Rearranging $F = m \times a$ gives
$$a = \frac{F}{m} = \frac{40 \, N}{5.0 \, kg} = 8.0 \, m/s^2$$

Key points

- The greater the resultant force on an object, the greater the object's acceleration.
- The greater the mass of an object, the smaller its acceleration for a given force.
- The resultant force acting on an object is $F = m \, a$
- Ⓗ The inertia of an object is its tendency to stay at rest or in uniform motion.

P10.2 Weight and terminal velocity

Learning objectives

After this topic, you should know:

- the difference between mass and weight
- about the motion of a falling object acted on only by gravity
- what terminal velocity means
- what can be said about the resultant force acting on an object that is falling at terminal velocity.

If you release an object above the ground, it falls because of its weight (i.e., the force acting on the object due to gravity).

If no other forces act on it, the resultant force on it is its weight. The object is said to be falling freely. It accelerates downwards at a constant acceleration of $9.8 \, \text{m/s}^2$. This is the acceleration due to gravity (or the acceleration of free fall) and is represented by the symbol g. For example, if you release a 1 kg object above the ground:

- the gravitational force on it is 9.8 N, and
- its acceleration $\left(= \dfrac{\text{force}}{\text{mass}} = \dfrac{9.8 \, \text{N}}{1 \, \text{kg}} \right) = 9.8 \, \text{m/s}^2$.

Mass and weight

Your weight is caused by the gravitational force of attraction between you and the Earth. This force is very slightly weaker at the equator than at the poles, so at the equator you will weigh slightly less than at the poles. However, your mass will be the same no matter where you are.

- The **weight** of an object is the force acting on it due to gravity. Weight is measured in newtons, N.
- The **mass** of an object depends on the quantity of matter in it. Mass is measured in kilograms, kg.

You can measure the weight of an object by using a newton-meter. The weight of an object:

- of mass 1 kg is 9.8 N
- of mass 5 kg is 50 N.

The gravitational force on a 1 kg object is the **gravitational field strength** at the place where the object is. Gravitational field strength is measured in Newtons per kilogram (N/kg). The Earth's gravitational field strength at its surface is about 9.8 N/kg.

If we know the mass of an object, we can calculate the force due to gravity which acts on it (i.e., its weight) using the equation:

weight, W	=	**mass, m**	×	**gravitational field strength, g**
(newtons, N)		(kilograms, kg)		(newtons per kilogram, N/kg)

Terminal velocity

If an object falls in a fluid, the fluid drags on the object because of friction between the fluid and the surface of the moving object. This frictional force increases with speed. At any instant, the resultant force on the object is its weight minus the frictional force on it.

- The acceleration of the object decreases as it falls. This is because the frictional force increases as it speeds up. So the resultant force on it decreases and therefore its acceleration decreases.

Worked example

Calculate the weight in newtons of a person of mass 55 kg.

Solution

Weight = mass × gravitational field strength

$= 55 \, \text{kg} \times 9.8 \, \text{N/kg} = \mathbf{540 \, N}$

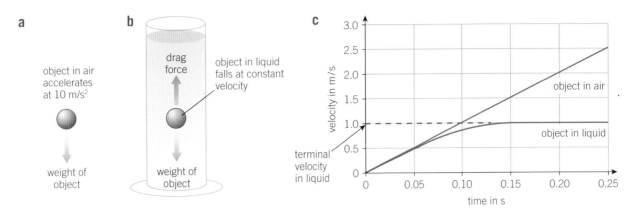

Figure 1 *Falling objects.* **a** *Falling in air,* **b** *falling in a liquid, and* **c** *velocity–time graph for* **a** *and* **b**

- The object reaches a constant velocity when the frictional force on it is equal and opposite to its weight. This velocity is called its **terminal velocity**. The resultant force is then zero, so its acceleration is zero.

When an object moves through the air (instead of water), the frictional force is called air resistance. This is not shown in Figure 1a because the air resistance on the object is much smaller than the frictional force on it when it is falling in the liquid in (Figure 1b). The object in Figure 1a would need to fall much further through the air before it reaches a constant velocity.

1 An object is released in a fluid. Describe:
   a the resultant force on it initially [1 mark]
   b the weight of the object and the frictional force on it before it reaches its terminal velocity [1 mark]
   c its acceleration after it reaches its terminal velocity [1 mark]
   d the resultant force on it when it moves at its terminal velocity. [1 mark]

2 The gravitational field strength at the surface of the Earth is 9.8 N/kg. For the Moon, it is 1.6 N/kg.
   a Calculate the weight of a person of mass 50 kg on the Earth. [1 mark]
   b Calculate the weight of the same person if she was on the Moon. [1 mark]
   c A lunar vehicle weighs 300 N on the Earth. Calculate its weight on the Moon. [2 marks]

3 A parachutist of mass 70 kg supported by a parachute of mass 20 kg reaches a constant speed.
   a Explain why the parachutist reaches a constant speed. [4 marks]

   b Calculate:
   i the total weight of the parachutist and the parachute [1 mark]
   ii the size and direction of the force of air resistance on the parachute when the parachutist falls at constant speed. [1 mark]

4 **a** Use Figure 1 to determine the acceleration of the object in liquid 0.10 s after it was released. [2 marks]
   b Show that the ratio of the drag force to the weight at 0.1 s is about 0.5. [3 marks]

Study tip

When the upward force acting on an object falling in a fluid balances the downward force, the object continues at a constant speed – it doesn't stop!

Key points

- The weight of an object is the force acting on the object due to gravity. Its mass is the quantity of matter in the object.
- An object acted on only by gravity accelerates at about 10 m/s².
- The terminal velocity of an object is the velocity it eventually reaches when it is falling. The weight of the object is then equal to the frictional force on the object.
- When an object is moving at terminal velocity, the resultant force on it is zero.

P10.3 Forces and braking

Learning objectives

After this topic, you should know:

- the forces that oppose the driving force of a vehicle
- what the stopping distance of a vehicle depends on
- what can increase the stopping distance of a vehicle
- how to estimate the braking force of a vehicle.

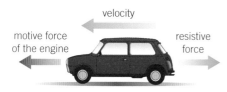

Figure 1 *Constant velocity*

Forces on the road

For any car travelling at constant velocity, the resultant force on it is zero. This is because the driving force of its engine is balanced by the resistive forces, which are mostly caused by air resistance.

A car driver uses the accelerator pedal to vary the driving force of the engine.

The braking force needed to stop a vehicle in a given distance depends on:

- the speed of the vehicle when the brakes are first applied
- the mass of the vehicle.

You can see this using the equation 'resultant force = mass × acceleration', which you first met in Topic P10.1, where the braking force is the resultant force.

- The greater the speed, the greater the deceleration needed to stop the vehicle within a given distance. So the braking force required to stop a car travelling at a high speed must be greater than the braking force required to stop a car travelling at a low speed, within the same distance.
- The greater the mass, the greater the braking force needed for a given deceleration.

Stopping distances

Driving tests always ask about **stopping distances**. This is the shortest distance a vehicle can safely stop in, and is in two parts:

1 The **thinking distance** – the distance travelled by the vehicle in the time it takes the driver to react (i.e., during the driver's reaction time). Because the car moves at constant speed during the reaction time, the thinking distance is equal to the speed × the reaction time. This shows that the thinking distance is proportional to the speed.

2 The **braking distance** – the distance travelled by the vehicle during the time the braking force acts.

stopping distance = thinking distance + braking distance

Figure 2 shows the stopping distance for a vehicle on a dry flat road travelling at different speeds. Check for yourself that the stopping distance at 31 m/s (70 miles per hour) is 96 m.

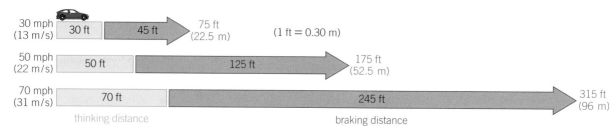

Figure 2 *Stopping distances*

Factors affecting stopping distances

1 Tiredness, alcohol, and drugs affect the brain and increase reaction times. Distractions like using a mobile phone also increase reaction time and cause serious accidents. All these factors increase the thinking distance (because thinking distance = speed × reaction time). So the stopping distance is greater.

2 The faster a vehicle is travelling, the further it travels before it stops. This is because the thinking distance and the braking distance both increase with increased speed.

3 In adverse road conditions, for example on wet or icy roads, drivers have to brake with less force to avoid skidding. So stopping distances are greater in poor weather and road conditions.

4 Poorly maintained vehicles, for example with worn brakes or tyres, take longer to stop because the brakes and tyres are less effective.

Road vehicles and forces

The deceleration of a road vehicle depends on the friction between the road and the car tyres. To avoid skidding on a dry flat road, the deceleration should be no more than about 6 m/s^2. Check this yourself using the equation $a = \dfrac{v^2 - u^2}{2s}$ and data from Figure 2.

Vehicle masses range from about 1000 kg for a car to about 38 000 kg for a heavy truck. So braking forces vary from about 6 kN for a car to about 250 kN for a heavy truck. The same range of forces are needed to accelerate a road vehicle.

1 For each of the following factors, identify which distance of a vehicle (thinking distance or breaking distance) is affected:
   a the road surface   [1 mark]
   b the tiredness of the driver   [1 mark]
   c poorly maintained brakes.   [1 mark]

2 **a** Use the chart in Figure 2 to calculate, in metres, the effect of an increase from 13 m/s (30 mph) to 22 m/s (50 mph) on the following:
   i the thinking distance [1 mark]   **ii** the braking distance [1 mark]
   iii the stopping distance.   [1 mark]
   b A driver has a reaction time of 0.8 s. Calculate the change in her thinking distance if she travels at 15 m/s instead of 30 m/s.   [2 marks]

3 **a** When the speed of a car is doubled:
   i explain why the thinking distance of the driver is doubled, assuming that the driver's reaction time is unchanged   [2 marks]
   ii explain why the braking distance is more than doubled.   [2 marks]
   b A student thinks that braking distance is proportional to the square of the speed. Use the chart in Figure 2 to decide whether or not this is a valid claim.   [3 marks]

4 A car of mass 1500 kg moving at 31 m/s decelerates and stops in a distance of 75 m.
   a ⒣ Calculate the deceleration of the vehicle.   [2 marks]
   b Estimate the braking force on the car.   [2 marks]

Deceleration

The deceleration a of a vehicle can be calculated using the equation $v^2 = u^2 + 2as$, where s is the distance travelled, u is the initial speed, and v is the final speed. You first met this equation in Topic P9.4. Rearranging the equation with $v = 0$ for a vehicle that stops gives $s = \dfrac{-u^2}{2a}$ or $a = \dfrac{-u^2}{2s}$.

- $s = \dfrac{-u^2}{2a}$ shows that s is proportional to u^2 for constant deceleration.

- $a = \dfrac{-u^2}{2s}$ can be used to calculate the deceleration when the values for u and s are given.

Braking force

To work out the braking force, use the equation $F = ma$ with given values or estimates for m and a.

Synoptic links

You learnt about friction in Topic P8.3.

Key points

- Friction and air resistance oppose the driving force of a vehicle.
- The stopping distance of a vehicle depends on the thinking distance and the braking distance.
- High speed, poor weather conditions, and poor vehicle maintenance all increase the braking distance. Poor reaction time (due to tiredness, alcohol, drugs, or using a mobile phone) and high speed both increase the thinking distance.
- $F = ma$ gives the braking force of a vehicle.

P10.4 Momentum

Learning objectives

After this topic, you should know:

- how to calculate momentum
- the unit of momentum
- what momentum means for a closed system
- that two objects that push each other apart move away with equal and opposite momentum.

Figure 1 *A contact sport*

Worked example

Calculate the momentum of a sprinter of mass 50 kg running at a velocity of 10 m/s.

Solution

$$\text{Momentum} = \text{mass} \times \text{velocity}$$
$$= 50\,\text{kg} \times 10\,\text{m/s}$$
$$= \textbf{500 kg m/s}$$

Synoptic links

Momentum is a vector quantity because it has size and direction. See Topic P8.1.

Momentum is important to anyone who plays a contact sport. In a game of rugby, a player with a lot of momentum is very difficult to stop.

momentum of a moving object, *p* (kg m/s) = mass, *m* (kg) × velocity, *v* (m/s)

Momentum has both a size and a direction so it is a vector quantity.

Investigating collisions

When two objects collide, the momentum of each object changes. Figure 2 shows how to use a computer and a motion sensor to investigate a collision between two trolleys.

Trolley A is given a push so that it collides with a stationary trolley B. The two trolleys stick together after the collision. The computer gives the velocity of A before the collision and the velocity of both trolleys afterwards.

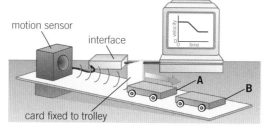

Figure 2 *Investigating collisions*

1   For two trolleys of the same mass, the velocity of trolley A is halved by the impact. The combined mass after the collision is twice the moving mass before the collision. So the momentum (= mass × velocity) after the collision is the same as the momentum before the collision.

2   For a single trolley pushed into a double trolley, the velocity of A is reduced to one-third. The combined mass after the collision is three times the initial mass. So in this test as well, the momentum after the collision is the same as the momentum before the collision.

In both tests, the total momentum is unchanged (i.e., is conserved) by the collision. This is an example of the **conservation of momentum**. It applies to any system of objects as long as the system is a closed system, which means that no resultant force acts on it.

In general, the law of conservation of momentum says that:

In a closed system, the total momentum before an event is equal to the total momentum after the event.

You can use this law to predict what happens whenever objects collide or push each other apart in an 'explosion'. Momentum is conserved in any collision or explosion as long as no external forces act on the objects.

If you are a skateboarder, you will know that your skateboard can shoot away from you when you jump off it. Its momentum is in the opposite direction to your own momentum. What can you say about the momentum of objects when they fly apart from each other?

Investigating a controlled explosion

Figure 3 shows a controlled explosion using trolleys. When the trigger rod is tapped, a bolt springs out, and the trolleys recoil (spring back) from each other.

You can also test what happens if one of the trolleys is a 'double trolley', as shown in Figure 4.

Using trial and error, you can place blocks on the runway so that the trolleys reach them at the same time. This lets you compare the speeds of the trolleys. Some results are shown in Figure 4.

Safety: Protect yourself and the bench from falling objects.

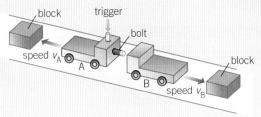

Figure 3 *Investigating explosions*

Figure 4 *Using different masses*

- Two single trolleys travel equal distances in the same time. This shows that they recoil at equal speeds.
- A double trolley travels only half the distance that a single trolley does. Its speed is half that of the single trolley.

In each test:

1 the mass of the trolley × the speed of the trolley is the same, and

2 they recoil in opposite directions.

So momentum has size and direction. The results show that the trolleys recoil with equal and opposite momentum.

1 **a** Define momentum and give its unit. [1 mark]
   b Calculate the momentum of a 40 kg person running at 6 m/s. [1 mark]

2 **a** Calculate the momentum of an 80 kg rugby player running at a velocity of 5 m/s. [1 mark]
   b An 800 kg car moves with the same momentum as the rugby player in **a**. Calculate the velocity of the car. [1 mark]
   c Calculate the velocity of a 0.40 kg ball that has the same momentum as the rugby player in **a**. [1 mark]

3 A 60 kg skater and an 80 kg skater standing in the middle of an ice rink push each other away (Figure 5).
   Describe:
   a the force they exert on each other when they push apart
   b the momentum each skater has just after they separate
   c each of their velocities just after they separate
   d their total momentum just after they separate. [5 marks]

4 In Question **3**, the 60 kg skater moves away at 2.0 m/s. Calculate:
   a her momentum [1 mark]
   b the velocity of the other skater. [3 marks]

Figure 5 *See Summary questions 3 and 4*

Key points

- The momentum of a moving object is $p = mv$
- The unit of momentum is kg m/s.
- A closed system is a system in which the total momentum before an event is the same as the total momentum after the event. This is called conservation of momentum.

P10.5 Forces and elasticity

Learning objectives

After this topic, you should know:

- what is meant when an object is called elastic
- how to measure the extension of an object when it is stretched
- how the extension of a spring changes with the force applied to it
- what is meant by the limit of proportionality of a spring.

Squash players know that hitting a squash ball changes the ball's shape briefly. The shape of an object can be changed by stretching, bending, twisting, or compressing it. A squash ball is **elastic** because it goes back to its original shape. The ball is said to have been elastically deformed.

A rubber band is also elastic because it returns to its original length after it is stretched and then released. Rubber is an example of an elastic material. An object such as a polythene bag does not return to its original shape after being deformed and so is said to have been inelastically deformed.

An object is elastic if it returns to its original shape when the forces deforming it are removed.

Table 1 *Weight versus length measurements for a rubber strip*

Weight in N	Length in mm	Extension in mm
0.0	120	0
1.0	152	32
2.0	190	70
3.0	250	
4.0		

Stretch tests

You can investigate how easily a material or a spring stretches by hanging weights from it (Figure 1).

- The spring to be tested is clamped at its upper end. An empty weight hanger is attached to the spring to keep it straight.
- The length of the spring is measured using a metre ruler. This is its original length.
- The weight hung from the spring is increased by adding weights one at a time. The spring stretches each time more weight is hung from it.
- The length of the spring is measured each time a weight is added. The spring should be measured from the same points each time to ensure accurate results. The total weight added and the total length of the spring are recorded in a table.

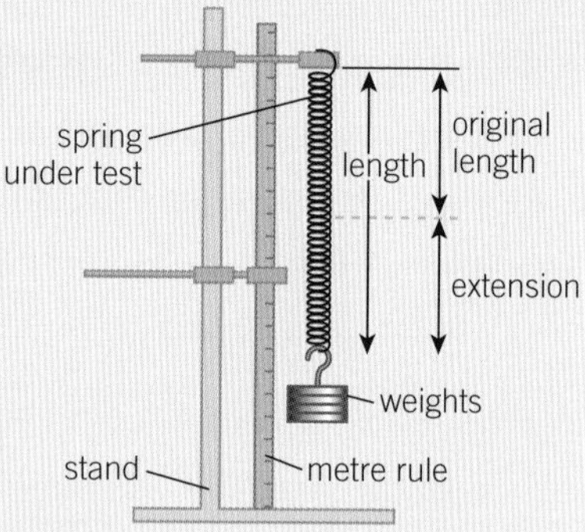

Safety: Clamp the stand to the bench and take care with falling weights. Wear eye protection.

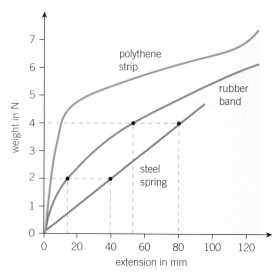

Figure 1 *Investigating stretching*

The increase of length from the original is called the **extension**. This is calculated each time a weight is added and recorded, as shown in Table 1.

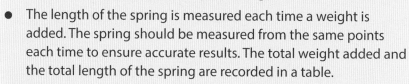

extension of the strip of material or spring at any stage = length at the stage – original length

The measurements can be plotted on a graph of weight on the vertical *y*-axis against extension on the *x*-axis. Figure 2 shows the results for strips of different materials and a steel spring plotted on the same axes.

The steel spring gives a straight line through the origin. This shows that the weight hung on the steel spring is **directly proportional** to the extension of the spring. For example, doubling the weight from 2.0 N to 4.0 N doubles the extension of the steel spring.

Figure 2 *Extension versus hung weight for different materials*

Hooke's law

In the tests above, the extension of a steel spring is directly proportional to the force applied to it. You can use the graph to predict what the extension would be for any given force. But if the force is too big, the spring stretches more than predicted. This is because the spring has been stretched beyond its **limit of proportionality**.

The extension of a spring is directly proportional to the force applied, as long as its limit of proportionality is not exceeded.

The above statement is known as Hooke's law. If the extension of any stretched object or material is directly proportional to the stretching force, the object or material is said to obey Hooke's law.

1 The lines on the graph in Figure 2 show that rubber and polythene have a low limit of proportionality. Beyond this limit, they do not obey Hooke's law. The relationship between their extension and the force applied is non-linear. A steel spring has a much higher limit of proportionality. Its relationship to the force applied stays linear for much longer.

2 Hooke's law can be written as an equation:

force applied, F =   spring constant, k   × extension, e
   (newtons, N)   (newtons per metre, N/m)   (metres, m)

The spring constant is equal to the force per unit extension needed to extend the spring, assuming that its limit of proportionality is not reached. The stiffer a spring is, the greater its spring constant.

3 Hooke's law also applies to an object when it is compressed. In this situation, the change of length is a compression, *not* an extension.

1 What is meant by:
   a the limit of proportionality of a spring   [1 mark]
   b the spring constant of a spring   [1 mark]
   c the extension of a stretched spring.   [1 mark]
2 a Describe what happens to a strip of polythene if it is stretched beyond its limit of proportionality.   [1 mark]
   b Compare the result of stretching then releasing a rubber band with the result of stretching a strip of polythene.   [1 mark]
3 a Look at Figure 2. When the weight is 4.0 N, calculate the extension of:
      i   the spring   [1 mark]
      ii   the rubber band   [1 mark]
      iii the polythene strip.   [1 mark]
   b i Calculate the extension of the spring when the weight is 3.0 N.   [1 mark]
      ii Calculate the spring constant of the spring.   [2 marks]
4 a Write Hooke's law.   [1 mark]
   b A spring has a spring constant of 25 N/m.
      i   Calculate how much force is needed to make the spring extend by 0.10 m.   [2 marks]
      ii Calculate the extension of the spring when it hangs vertically from a fixed point and supports a 5.0 N weight at its lower end. [2 marks]

Using maths

You can write the word equation for Hooke's law using symbols as

$$F = k \times e$$

where

F = force in newtons, N

k = the spring constant in newtons per metre, N/m

e = extension in metres, m.
To determine the spring constant of a spring from a set of force and extension measurements, plot the force F on the y-axis (and the extension e on the x-axis). The gradient of the straight line you obtain is the spring constant k.

Synoptic link

You learnt about elastic potential energy stores, including the equation to calculate them, in Topic P1.5.

Key points

- An object is called elastic if it returns to its original shape after removing the force deforming it.
- The extension is the difference between the length of the object and its original length.
- The extension of a spring is directly proportional to the force applied to it, as long as the limit of proportionality is not exceeded. This relationship is linear.
- Beyond the limit of proportionality, the extension of a spring is no longer proportional to the applied force. This relationship becomes non-linear.

P10 Force and motion

Summary questions

1 a Give the reason why the stopping distance of a car is increased if:
  i the road is wet instead of dry [1 mark]
  ii the driver is tired instead of alert. [1 mark]
 b A driver travelling at 18 m/s takes 0.7 s to react when a dog walks into the road 40 m ahead. The braking distance for the car at this speed is 24 m.
  i Calculate the distance travelled by the car in the time it takes the driver to react. [2 marks]
  ii Calculate how far in front of the dog the car stops. [2 marks]
  iii The total mass of the car and its contents is 1200 kg. Calculate the car's deceleration when the brakes are applied, and so calculate the braking force. **H**
   [3 marks]

2 A space vehicle of mass 200 kg rests on its four wheels on a flat area of the lunar surface. The gravitational field strength at the surface of the Moon is 1.6 N/kg.
 a Calculate the weight of the space vehicle on the lunar surface. [2 marks]
 b Calculate the force that each wheel exerts on the lunar surface. [1 mark]

3 a A racing cyclist accelerates at 5.0 m/s² when she starts from rest. The total mass of the cyclist and her bicycle is 45 kg. Calculate:
  i the resultant force that produces this acceleration [2 marks]
  ii the total weight of the cyclist and the bicycle. [2 marks]
 b Explain why she can reach a higher speed by crouching than by staying upright. [4 marks]

4 In a Hooke's law test on a spring, the following results were obtained.

Weight in newtons	Length in millimetres	Extension in millimetres
0.0	245	0
1.0	285	40
2.0	324	
3.0	366	
4.0	405	
5.0	446	
6.0	484	

a Copy and complete the third column of the table. [1 mark]
 b Plot a graph of the extension on the vertical axis against the weight on the horizontal axis. [3 marks]
 c If a weight of 7.0 N is suspended on the spring, work out what the extension of the spring would be. [2 marks]
 d i Calculate the spring constant of the spring. [2 marks]
  ii An object suspended on the spring gives an extension of 140 mm. Calculate the weight of the object. [2 marks]

5 **H** A car of mass 1500 kg is moving at a speed of 30 m/s on a horizontal road when the driver applies the brakes and the car stops 12 s later.

30 m/s

Figure 1

a i Calculate the initial momentum of the car before the brakes are applied. [1 mark]
  ii Calculate the braking force. [2 marks]
 b Describe how the momentum of the car changes when the brakes are applied. [1 mark]
 c Evaluate the effect on the motion of the car if the brakes had been applied with much greater force. [2 marks]

6 **H** When a stationary football of mass 0.44 kg was kicked, its velocity increased to 19 m/s as a result of the impact.
 a Calculate the gain of momentum of the football due to the impact. [2 marks]
 b The impact lasted 0.0384 s. Calculate the acceleration of the ball due to the impact force and show that the impact force was 220 N. [3 marks]

7 A skateboarder on a stationary skateboard jumped forwards off the skateboard. As a result, the skateboard moved away backwards. Use the idea of conservation of momentum to explain why the skateboard moved backwards. [4 marks]

Practice questions

01 A student uses a reaction time tester linked to digital timer. The student started the test by pressing a button. After a random amount of time time has elapsed an LED came on and the student pressed the button again. The digital timer recorded the time between the LED coming on and the student pressing the button.

01.1 Suggest one reason why this reaction time test is better than using the dropping ruler method.
[1 mark]

01.2 Name the independent variable in the test. [1 mark]

01.3 The stopping distance of a car is thinking distance + braking distance.
Which two factors increase the thinking distance? Choose the correct answers.

| tiredness | alcohol | brakes | tyres |

[2 marks]

01.4 The acceleration of a car on a straight road is 3.28 m/s². The mass of the car is 1200 kg. Calculate the resultant force on the car. Give your answer correct to 2 significant figures. [2 marks]

01.5 Complete the sentence using the correct word from the box.

| acceleration | speed | deceleration |

The greater the braking force on a car the greater the _____ of the car. [1 mark]

01.6 Describe the dangers of a very large deceleration on the motion of a car. [2 marks]

02 A student investigated how the pulling force on a trolley affected the acceleration of the trolley between two points. The apparatus he used is shown in **Figure 1**.

Figure 1

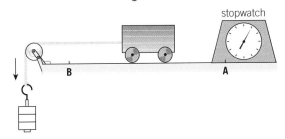

The trolley was held at position **A** and weights were suspended by a hook. The trolley was then released and the time taken to travel between **A** and **B** was recorded. The test was repeated using weights with different values.

02.1 Name the factors that should remain constant (the control variables) during the investigation. [3 marks]

02.2 The results of the investigation were recorded in **Table 1**.

Table 1

Force in N	100	200	300	400	500
Acceleration in m/s²	0.25	0.50	0.65	1.00	1.25

Plot a graph of the results. [3 marks]

02.3 Identify which one of the results is anomalous and suggest a reason for this anomalous result. [2 marks]

02.4 The student predicted that if the force was doubled the acceleration would also be doubled. Describe how the graph can be used to confirm his prediction. [2 marks]

02.5 A stopwatch was used to measure the time taken for the trolley to travel between points A and B. Another student suggested that light gates attached to a data logger would improve the investigation. Give two advantages of using light gates and a data logger to measure the acceleration. [2 marks]

03 A baby bouncer consists of a harness seat for a toddler, attached to a spring. The idea is for the baby to hang in the seat with his feet just touching the floor, so that a good push up will get the baby bouncing.

Figure 2

03.1 The spring is tested in the laboratory. When a force of 10 N is suspended from the spring, the extension of the spring is 0.02 m. Calculate the spring constant of the spring and give the unit. [3 marks]

03.2 The mass of the baby is 9 kg. Calculate the weight of the baby.
gravitational field strength = 9.8 N/kg. [2 marks]

03.3 The baby is placed unsuspended in the harness. His feet are 0.68 m from the lower end of the spring. Calculate the height above the floor the lower end of the stretched spring needs be if his feet are just to touch the floor. Use the information from question **03.2**. [3 marks]

03.4 The maximum permissible mass of the baby on the bouncer is 15 kg. Explain why it is important to test the spring beyond this maximum figure. [2 marks]

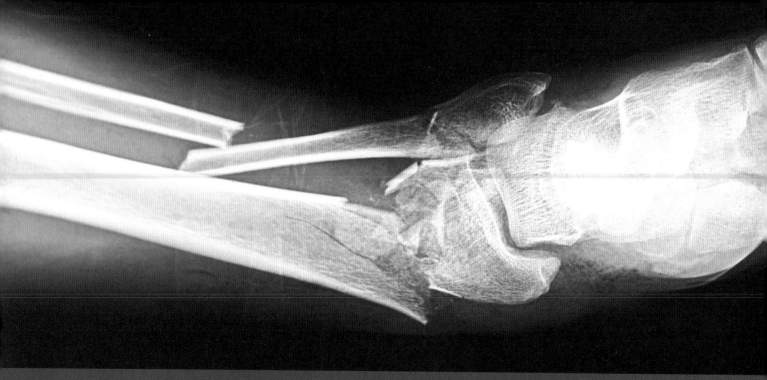

4 Waves and electromagnetism

When you speak into a mobile phone, you create sound waves that carry information. These waves are detected by a microphone that produces electrical waves in the phone circuits. Your phone then sends out radio waves that carry the information to your mobile phone network and then to the person you are calling. Medical doctors use radio waves in scanners to obtain 3D images of organs. They also use X-rays to visualise objects inside the body.

In this section, you will learn about waves and their properties, and the many ways they are used. You will also learn about magnetic fields and how we use them to lift objects or make them move.

Key questions

- How do we measure waves and how fast do they travel?

- What are electromagnetic waves and how do they differ from sound waves?

- How do waves carry information?

Making connections

- Electric motors are using in many appliances such as washing machines, hairdryers, and electric drills. In this section you will learn about electric motors and the magnetic effects of electric currents. When you study these topics, look back to **P1 Conservation and dissipation of energy** to remind yourself about some of the electrical appliances we use.

I already know...

The top of a water wave is called a crest and the bottom is called a trough.

Light travels much faster than sound and can travel through space whereas sound cannot.

The spectrum of white light is continuous from red to orange, yellow to green, and blue to violet.

There are different types of waves, such as sound waves and electromagnetic waves, but they all have common properties such as refraction.

A magnet lines up with the Earth's magnetic field.

An electric motor is used to turn objects.

I will learn...

How the wavelength of a wave depends on its speed and its frequency.

How to measure the speed of sound waves in air and in a solid.

How electromagnetic waves carry information and how they are used to form images.

What we mean by refraction of waves when they cross a boundary between different substances.

How the strength of a magnetic field is measured and what a solenoid is.

How an electric motor works.

Required Practicals

Practical		Topic
20	Investigating plane waves in a ripple tank and waves in a solid.	P11.4
21	Investigating infrared radiation	P12.2

Learning objectives

After this topic, you should know:

- what waves can be used for
- what transverse waves are
- what longitudinal waves are
- which types of waves are transverse and which are longitudinal.

Figure 1 *Waves in water are examples of mechanical waves*

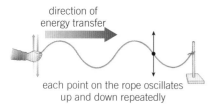

direction of energy transfer

each point on the rope oscillates up and down repeatedly

Figure 2 *Transverse waves*

Waves transfer energy without transferring matter. You can also use waves to transfer information, for example when you use a mobile phone or listen to the radio.

There are different types of waves. These include:

- sound waves, water waves, waves on springs and ropes, and seismic waves produced by earthquakes. These are all examples of **mechanical waves**, which are vibrations that travel through a medium (a substance).

- light waves, radio waves, and microwaves. These are all examples of **electromagnetic waves**, which can all travel through a vacuum at the same speed of 300 000 kilometres per second. No medium is needed.

Observing mechanical waves

Figure 2 shows how you can make waves on a rope by moving one end up and down.

Tie a ribbon to the middle of the rope. Move one end of the rope up and down. You will see that the waves move along the rope but the ribbon doesn't move along the rope – it just moves up and down. This type of wave is known as a **transverse wave**. It is said that the ribbon **vibrates** or **oscillates**. This means that it moves repeatedly between two positions. When the ribbon is at the top of a wave, it is said to be at the peak (or crest) of the wave.

Repeat the test with the slinky. You should observe the same effects if you move one end of the slinky up and down.

However, if you push and pull the end of the slinky as shown in Figure 3, you will see a different type of wave, known as a **longitudinal wave**. Notice that there are areas of **compression** (coils squashed together) and areas of **rarefaction** (coils spread further apart) moving along the slinky.

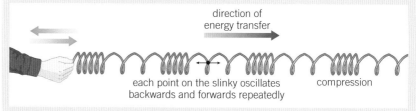

direction of energy transfer

each point on the slinky oscillates backwards and forwards repeatedly

compression

Figure 3 *Making longitudinal waves on a slinky*

- Describe how the ribbon moves when you send longitudinal waves along the slinky.

Safety: Handle the slinky spring carefully.

More wave tests

When waves travel through a substance, the substance itself doesn't travel. You can see this with waves on a rope and by observing:

- waves spreading out on water after a small object is dropped in the water. The waves travel across the surface but the water does not travel away from the object.

- a tuning fork vibrating so that it makes sound waves travel through the air away from the tuning fork. The air itself doesn't travel away from the vibrating object – if it did, a vacuum would be created.

Transverse waves

Imagine sending waves along a rope that has a white spot painted on it. You would see the spot move up and down without moving along the rope. In other words, the spot would oscillate perpendicular (at right angles) to the direction in which the waves are moving. The waves on a rope and the ripples on the surface of water are called transverse waves because the vibrations (called oscillations) move up and down or from side to side. All electromagnetic waves are transverse waves.

The oscillations of a transverse wave are perpendicular to the direction in which the waves transfer energy.

Longitudinal waves

The slinky spring in Figure 3 is useful to demonstrate how sound waves travel. When one end of the slinky is pushed in and out repeatedly, vibrations travel along the spring. These oscillations are parallel to the direction in which the waves transfer energy. Waves that travel in this way are called longitudinal waves.

Sound waves travelling through air are longitudinal waves. When an object vibrates in air, it makes the air around it vibrate as it pushes and pulls on the air. The oscillations (compressions and rarefactions) that travel through the air are sound waves. The oscillations are along the direction in which the wave travels.

The oscillations of a longitudinal wave are parallel to the direction in which the waves transfer energy.

Mechanical waves can be transverse or longitudinal.

1   **a** What is the difference between a longitudinal wave and a transverse wave? [1 mark]
    b Give *one* example of:
       i a transverse wave [1 mark]   **ii** a longitudinal wave. [1 mark]
    c When a sound wave passes through air, describe what happens to the air particles at a point of compression. [1 mark]
2   A long rope with a knot tied in the middle lies straight along a smooth floor. A student picks up one end of the rope. This sends waves along the rope.
    a Determine whether the waves on the rope are transverse or longitudinal waves. [1 mark]
    b Describe:
       i the direction of energy transfer along the rope [1 mark]
       ii the movement of the knot. [1 mark]
3   Describe how to use a slinky spring to demonstrate to a friend the difference between longitudinal waves and transverse waves. [2 marks]
4   Describe and explain the motion of a small ball floating on a pond when waves travel across the pond. ⓘ [3 marks]

Synoptic links

You will learn more about electromagnetic waves in Topic P12.1.

Study tip

- Make sure you can know how to describe the difference between transverse waves and longitudinal waves.

- Remember that electromagnetic waves are transverse, and sound waves are longitudinal.

Key points

- Waves can be used to transfer energy and information.
- Transverse waves oscillate perpendicular to the direction of energy transfer of the waves. Ripples on the surface of water are transverse waves. So are all electromagnetic waves.
- Longitudinal waves oscillate parallel to the direction of energy transfer of the waves. Sound waves in air are longitudinal waves.
- Mechanical waves need a medium (a substance) to travel through. They can be transverse or longitudinal waves.

P11.2 The properties of waves

Learning objectives

After this topic, you should know:

- what is meant by the amplitude, frequency, and wavelength of a wave
- how the period of a wave is related to its frequency
- the relationship between the speed, wavelength, and frequency of a wave
- how to use the wave speed equation in calculations.

If you want to find out how much energy or information waves carry, you need to measure them. Figure 1 shows a snapshot of waves on a rope. The crests, or peaks, are at the top of the wave. The troughs are at the bottom. They are equally spaced.

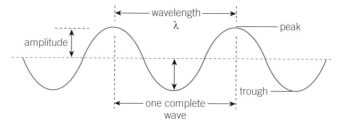

Figure 1 *Waves on a rope*

The **amplitude** of a wave is the maximum displacement of a point on the wave from its undisturbed position. For example, in Figure 1, this is the height of the wave crest (or the depth of the wave trough) from the middle.

The bigger the amplitude of the waves, the more energy the waves carry.

The **wavelength** of a wave is the distance from a point on the wave to the equivalent point on the adjacent wave. For example, in Figure 1, this is the distance from one wave crest to the next wave crest.

Frequency

If you made a video of the waves on the rope in Figure 1, you would see the waves moving steadily across the screen. The number of waves passing a fixed point every second is called the **frequency** of the waves.

The unit of frequency is the hertz (Hz). For the waves on the rope, one wave crest passing each second is equal to a frequency of 1 Hz.

The period of a wave is the time taken for each wave to pass a fixed point. For waves of frequency f, the period T is given by the equation:

$$\textbf{period, } \textbf{\textit{T}} \text{ (seconds, s)} = \frac{1}{\textbf{frequency, } \textbf{\textit{f}} \text{ (hertz, Hz)}}$$

Wave speed

Figure 2 shows a ripple tank, which is used to study water waves in controlled conditions. You can make straight waves by moving the long edge of a ruler up and down on the water surface in a ripple tank. Straight waves are called plane waves. The waves all move at the same speed and stay the same distance apart.

The **speed** of the waves is the distance travelled by each wave every second through a medium. Energy is transferred by the waves at this speed.

For waves of constant frequency, the speed of the waves depends on the frequency and the wavelength as follows:

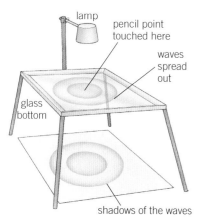

Figure 2 *The ripple tank*

wave speed, *v*   =   **frequency, *f***   ×   **wavelength, λ**
(metres per second, m/s)   (hertz, Hz)   (metres, m)

To understand what the wave speed equation means, look at Figure 3. The surfer is riding on the crest of some unusually fast waves.

Suppose the frequency of the waves is 3 Hz and the wavelength of the waves is 4.0 m.

- At this frequency, three wave crests pass a fixed point once every second (because the frequency is 3 Hz).

- The surfer therefore moves forward a distance of three wavelengths every second, which is 3 × 4.0 m = 12 m.

So the speed of the surfer is 12 m/s.

This speed is equal to the frequency × the wavelength of the waves.

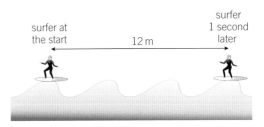

Figure 3 *Surfing*

Measuring the speed of sound in air

You need two people for this. You and a friend need to stand on opposite sides of a field at a measured distance apart. You should be as far apart as possible but within sight of each other.

If your friend bangs two cymbals together, you will see them crash together straightway, but you won't hear them straightaway. The crashing sound will be delayed because sound travels much slower than light. Use a stopwatch to time the interval between seeing the impact and hearing the sound. Repeat the test several times to get an average value of the time interval.

Calculate the speed of sound in air using the equation

$$\text{speed} = \frac{\text{distance}}{\text{time taken}}$$

1 Explain what is meant by the frequency of a wave.   [1 mark]

2 Figure 4 shows a wave travelling from left to right along a rope.
   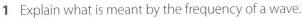
   a Copy the figure and mark on your diagram:
      i one wavelength   **Figure 4** [1 mark]
      ii the amplitude of the waves.   [1 mark]
   b Describe the motion of point P on the rope when the wave crest at P moves along by a distance of one wavelength.   [2 marks]

3 **a** A speedboat on a lake sends waves travelling across a lake at a frequency of 2.0 Hz and a wavelength of 3.0 m. Calculate the speed of the waves.   [2 marks]
   b If the waves had been produced at a frequency of 1.0 Hz and travelled at the speed calculated in **a**:
      i calculate what their wavelength would be   [2 marks]
      ii calculate the distance travelled by a wave crest in 60 s.[2 marks]

4 Sound waves in air travel at a speed of 340 m/s.
   a Calculate how far they travel in air in 5.0 s.   [2 marks]
   b Calculate their wavelength if their frequency is 3.0 kHz.   [2 marks]

Key points

- For any wave, its amplitude is the maximum displacement of a point on the wave from its undisturbed position, such as the height of the wave crest (or the depth of the wave trough) from the position at rest.
- For any wave, its frequency is the number of waves passing a point per second.
- The period of a wave = $\dfrac{1}{\text{frequency}}$
- For any wave, its wavelength is the distance from a point on the wave to the equivalent point on the next wave (e.g., from one wave trough to the next wave trough).
- The speed of a wave is $v = f \times \lambda$.

P11.3 Reflection and refraction

Learning objectives

After this topic, you should know:

- the patterns of reflection and refraction of plane waves in a ripple tank
- whether plane waves that cross a boundary between two different materials are refracted
- how the behaviour of waves can be used to explain reflection and refraction
- what can happen to a wave when it crosses a boundary between two different materials.

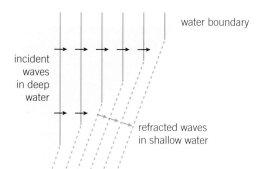

Wavefronts at a non-zero angle to the barrier

Incident wavefront

Reflected wavefront

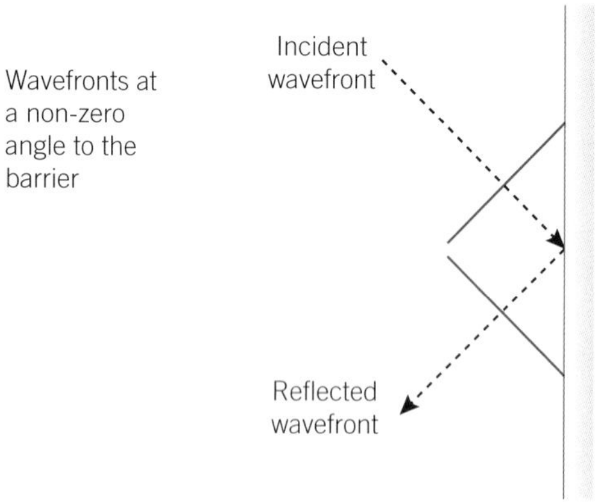

Figure 1 *Reflection of plane waves*

incident waves in deep water

water boundary

refracted waves in shallow water

Figure 2 *Refraction*

Investigating waves using a ripple tank

Reflection of waves can be investigated using the ripple tank. Each ripple is called a wavefront because it is the front of each wave as it travels across the water surface. Plane (i.e., straight) waves, produced by repeatedly dipping the long edge of a ruler in water, are directed at a metal barrier in the water. These waves are called the incident waves to distinguish them from the reflected waves. The incident waves are reflected by the barrier.

In Figure 1, the incident wavefront is not parallel to the barrier before or after reflection. The reflected wavefront moves away from the barrier at the same angle to the barrier as the incident wavefront.

A reflection test

Use a ruler to create and direct plane waves at a straight barrier (Figure 1). Find out if the reflected waves are always at the same angle to the barrier as the incident waves. You could align a second ruler with the reflected waves and measure the angle of each ruler to the barrier. Repeat the test for different angles.

Safety: Mop up any water spillages.

Refraction of waves is the change of the direction in which they are travelling when they cross a boundary between one medium and another medium. You can see this in a ripple tank when water waves cross a boundary between deep and shallow water. Plane waves directed at a non-zero angle to the boundary change direction as they cross the boundary, as shown in Figure 2.

Refraction tests

Use a vibrating beam to create plane waves continuously in a ripple tank containing a transparent plastic plate.

Arrange the plate so that the waves cross a boundary between the deep and shallow water. The water over the plate needs to be very shallow.

At a non-zero angle to a boundary. The waves change their speed and direction when they cross the boundary. Find out if plane waves change direction towards or away from the boundary when they cross from deep to shallow water.

Perpendicular to a boundary (at normal incidence). The waves cross the boundary without changing direction. However, their speed changes.

- Find out if the waves travel slower or faster when they cross the boundary.

Safety: Mop up any water spillages.

Explaining refraction

To explain how a wavefront moves forward, imagine that each tiny section creates a wavelet (a little wave) that travels forward (Figure 3). The wavelets move forward together to recreate the wavefront that created them.

Refraction

When plane waves cross a boundary at a non-zero angle to the boundary, each wavefront experiences a change in speed and direction.

In Figure 3, the wavefronts move more slowly after they have crossed the boundary. So the refracted wavefronts are closer together and are at a smaller angle to the boundary than the incident wavefronts.

The refracted waves and the incident waves have the same frequency, but they travel at a different speeds, so they have different wavelengths.

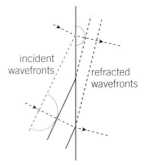

Figure 3 *Explaining refraction*

Materials and waves

When a wave is directed at a substance, some or all of the wave may be reflected at the surface. What happens is dependent on the wavelength of the wave and also on the substance (e.g., its surface). For example, microwaves are reflected by metal surfaces but they can pass through paper.

Of the waves that go into a substance, some or all of them may be absorbed by the substance. This would heat the substance because it would gain energy from the waves. For example, food is heated in microwave ovens because the microwaves are absorbed by the food.

As waves travel through a substance, the amplitude of the waves gradually decreases as the substance absorbs some of the waves' energy.

Waves that are not absorbed by the substance they are travelling through are transmitted by it. For example, light is mostly **transmitted** by ordinary glass, but is almost completely absorbed by darkened glass.

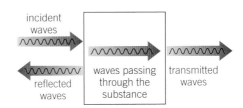

Figure 4 *Waves and substances*

1 When plane waves reflect from a straight barrier, describe the angle of each reflected wavefront to the barrier and the angle of each incident wavefront to the barrier. [1 mark]

2 Draw a diagram that shows plane waves passing from deep to shallow water at a non-zero angle to a straight boundary. Draw some refracted wavefronts, indicating their direction. [3 marks]

3 **a** Sea waves rolling up a sandy beach are not reflected. Explain why the sides of a ripple tank are sloped. [1 mark]

   b Describe what would happen if the sides of the ripple tank were vertical instead of sloped. [1 mark]

4 Sunglasses have lenses made of dark glass that reduce the amount of daylight entering your eyes. Design a test using a light meter and a lamp to find out if the two lenses in a pair of sunglasses are equally effective. [4 marks]

Key points

- Plane waves in a ripple tank are reflected from a straight barrier at the same angle to the barrier as the incident waves because their speed and wavelength do not change on reflection.
- Plane waves crossing a boundary between two different materials are refracted unless they cross the boundary at normal incidence.
- Refraction occurs at a boundary between two different materials because the speed and wavelength of the waves change at the boundary.
- At a boundary between two different materials, waves can be transmitted or absorbed.

P11.4 More about waves

Learning objectives

After this topic, you should know:

- what sound waves are
- how to investigate waves.

Figure 1 *Making sound waves – the buzzing of a bee is caused by the vibrations of its wings*

Sound waves are easy to produce. Your vocal cords vibrate and produce sound waves every time you speak. Any object vibrating in air makes the layers of air near the object vibrate, which make the layers of air next to them vibrate. The vibrating object pushes and pulls repeatedly on the air. This sends out vibrations of air in waves of compressions and rarefactions. When the waves reach your ears, they make your eardrums vibrate in and out so that you hear sound.

The vibrations travelling through the air are sound waves. The waves are longitudinal because the air particles vibrate (or oscillate) along the direction in which the waves transfer energy. The speed of sound waves in air increases with increasing temperature and is 340 m/s at 15 °C.

Investigating sound waves

You can use a loudspeaker to produce sound waves. Figure 2 shows how to do this using a signal generator.

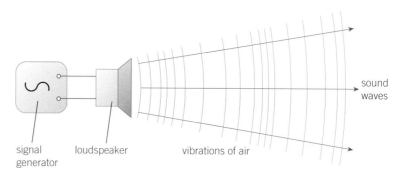

Figure 2 *Using a loudspeaker*

If you observe the loudspeaker closely, you can see it vibrating. It produces sound waves as it pushes the surrounding air backwards and forwards.

If you alter the frequency dial of the signal generator, you can change the frequency of the sound waves.

Sound waves cannot travel through a vacuum. You can test this by listening to an electric bell in a bell jar (Figure 3). As the air is pumped out of the bell jar, the ringing sound fades away.

A sound test

Sound waves reflect from a smooth hard surfaces, such as bare walls. If you clap your hands together in a large hall with bare walls, for example, a school gymnasium, you will hear the **echo** a short time later. The time delay is because the sound waves travel to the wall and back before you hear the echo.

If the distance d to the nearest wall is measured and the time delay t is also measured, the speed of sound in air can be calculated using the equation:

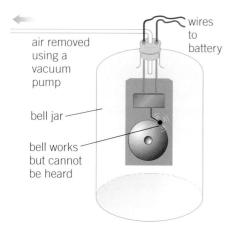

Figure 3 *A sound test – sound waves can't travel through a vacuum*

$$\text{speed, } s = \frac{\text{distance to the wall and back, } 2d}{\text{time delay, } t}$$

Investigating waves

Investigate waves on a stretched string using the apparatus shown in Figure 4. The oscillator sends waves along the string. You can adjust the frequency of the oscillator until there is a single loop on the string. Its length is half the length of one wavelength. The vibrating string sends out sound waves at the same frequency into the surrounding air.

- Note the frequency of the oscillator.
- Make suitable measurements to find the length, L, of a single loop and calculate the wavelength of the waves (= $2L$)
- Calculate the speed of the waves on the string using the equation: wave speed = frequency × wavelength
- Increase the frequency to obtain more loops on the string. Make more measurements to see if the wave speed is the same.

To measure the speed of the waves in a ripple tank (Figure 2, Topic P11.2), use a ruler to create plane waves that travel towards one end of the ripple tank.

- Use a stopwatch to measure the time it takes for a wave to travel from one end of the ripple tank to the other.
- Measure the distance the waves travel in this time.
- Use the equation speed = distance ÷ time to calculate the speed of the waves.

Observe the effect on the waves of moving the ruler up and down faster. More waves are produced every second and they are closer together.

- Determine whether the speed of the waves has changed.

Safety: Take care not to spill any liquids and, if you do, let your teacher know. You should also take care with hanging weights – clamp the stands to the bench and wear eye protection.

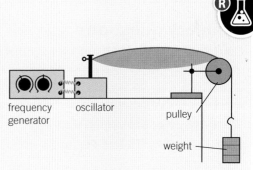

frequency generator   oscillator   pulley   weight

Figure 4 *Investigating waves on a string*

1 A lighting strike was heard by a student 4 seconds after he saw the lightning flash. Calculate the distance of the student to the point where the lightning strike occurred. Assume the speed of sound in air is 340 m/s. [2 marks]

2 A student standing at a fixed distance from a large bare wall clapped her hands repeatedly in time with the echoes she heard. Her friend timed 11 claps in 2.4 s. Estimate the shortest distance from the student to the wall.
Assume the speed of sound in air is 340 m/s. [3 marks]

3 A stretched string of length 1.24 m was made to oscillate as shown in Figure 4. A pattern of three equal loops was seen on the string when the oscillator vibrated at a frequency of 180 Hz.
a i Calculate the wavelength of the waves on the string. [2 marks]
ii Calculate the speed of the waves on the string. [2 marks]
b Explain why the vibrations of the string caused sound waves in the surrounding air. [3 marks]

Key points

- Sound waves are vibrations that travel through a medium (a substance).
- Sound waves cannot travel through a vacuum (e.g., in outer space).
- To investigate waves, use:
  - a ripple tank for water waves
  - a stretched string for waves in a solid
  - a signal generator and a loudspeaker for sound waves.

P11 Wave properties

Summary questions

1 a Figure 1 shows transverse waves on a string. Copy the diagram and label distances on it to show what is meant by:
   i the wavelength [1 mark]
   ii the amplitude of the waves. [1 mark]

Figure 1

b Describe the difference between a transverse wave and a longitudinal wave. [1 mark]

c Give **one** example of:
   i a transverse wave [1 mark]
   ii a longitudinal wave. [1 mark]

2 A speedboat on a lake creates waves that make a buoy bob up and down.
a The buoy bobs up and down three times in one minute. Calculate the frequency of the waves. [2 marks]
b The waves travel 24 metres in one minute. Calculate the speed of the waves in metres per second. [2 marks]
c Calculate the wavelength of the waves. [2 marks]

3 a When ripples travel across a water surface in a ripple tank, a small cork floating on the water surface bobs up and down without travelling across the tank. What does this tell you about the water in the ripple tank? [1 mark]

b Copy and complete Figure 2 to show the refraction of a straight wavefront at a straight boundary as the wavefront moves from deep to shallow water. [1 mark]

c Explain the change of direction of the wave in Figure 2 when it crosses the boundary. [3 marks]

Figure 2

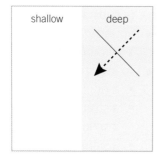

4 ⓗ a A loudspeaker is used to produce sound waves.
   i Describe how sound waves are created when an object in air vibrates. [3 marks]
   ii In terms of the amplitude of the sound waves, explain why the sound is fainter further away from the loudspeaker. [1 mark]

b A microphone is connected to an oscilloscope. Figure 3 shows the display on the screen of the oscilloscope when the microphone detects sound waves from a loudspeaker which is connected to a signal generator.

Figure 3

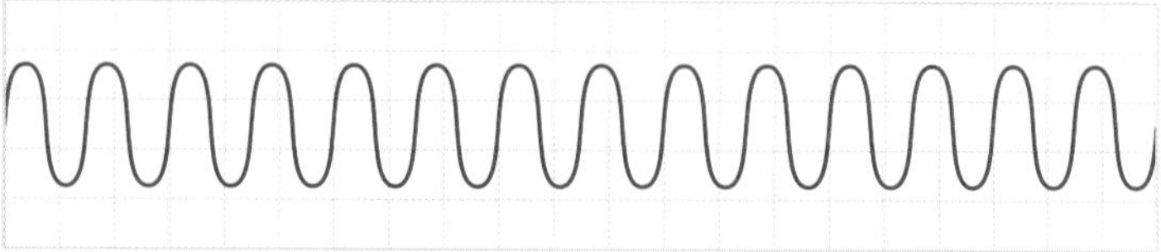

Describe how the waveform displayed on the oscilloscope screen changes if the sound from the loudspeaker is:
   i made louder [1 mark]
   ii reduced in pitch. [1 mark]

c Describe how you would use the arrangement to measure the upper frequency limit of a person's hearing. ✏ [5 marks]

5 A person is standing a certain distance from a flat side wall of a tall building. She claps her hands and hears an echo.
a Explain the cause of the echo. [2 marks]
b ⓗ She hears the sound 0.30 s after clapping her hands. Calculate how far she is from the nearest point of the wall. Assume the speed of sound in air is 340 m/s. [2 marks]

6 a A vibrating tuning fork sends out sound waves of frequency 256 Hz into the surrounding air. Calculate the frequency of these sound waves. Assume the speed of sound in air is 340 m/s. [2 marks]
b Give one similarity and one difference between sound waves in air and waves travelling across the water in a pond. [2 marks]

Practice questions

01.1 Sound travels in waves. Choose **two** statements that are correct.

> Sound waves are longitudinal.
> Sound waves are transverse.
> Sound wave oscillations are parallel to the direction of energy transfer.
> Sound wave oscillations are perpendicular to the direction of energy transfer.
> Sound waves travel through a vacuum.

[2 marks]

01.2 Copy Figure 1. Label a compression and a rarefaction on.

Figure 1

[2 marks]

01.3 The frequency of a sound wave is 440 oscillations every second. Calculate the time period of the tuning fork. [2 marks]

02 A student observes a large stone falling into a pool of water. A ripple of water is formed on the surface (Figure 2).

Figure 2

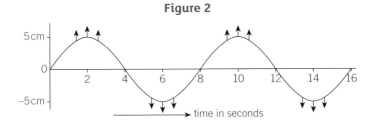

02.1 Determine whether the wave is a transverse or a longitudinal wave. [1 mark]

02.2 Determine the amplitude of the wave. [1 mark]

02.3 The student observes a plastic boat only moving up and down on the surface of the water. Explain why. [2 marks]

02.4 The wavelength of the wave is 28 cm. Calculate the speed of the wave and give the unit. [3 marks]

03 🄷 A group of students investigate the speed of sound in air. Two students stand a distance apart, with one holding an air horn and the other a stop watch.

- The first student drops his arm and at the same time sounds the horn.
- The second student starts the stop watch as the arm drops.
- The second student then stops the stop watch when she hears the horn.

03.1 Name the independent variable in the investigation. [1 mark]

03.2 🄷 Name the dependent variable in the investigation. [1 mark]

03.3 Give one reason why the second student should not stand too close to the first student to improve the results. [1 mark]

03.4 Suggest one safety precaution the first student should take. [1 mark]

03.5 Describe how errors can be minimised by repeating the test three times. [2 marks]

03.6 Describe how the speed of sound in air can be calculated from the results. [1 mark]

04 In an experiment to measure the speed of a water wave in a ripple tank, a student used a stopwatch to time how long a wave took to travel from one end of the tank to the other end and back. He repeated the measurement two more times and obtained the following timings:
2.3 s   2.5 s   2.5 s

04.1 Calculate the average time taken for the wave to travel from one end of the tank to the other and back. [2 marks]

04.2 The student used a metre ruler to measure the distance from one end of the ripple tank to the other end. She obtained a measurement of 58.5 cm. Calculate the speed of the water wave in cm/s. [2 marks]

04.3 Suggest how the accuracy of the timings could have been improved. [2 marks]

Learning objectives

After this topic, you should know:

- the parts of the electromagnetic spectrum
- the range of wavelengths within the electromagnetic spectrum that the human eye can detect
- how energy is transferred by electromagnetic waves
- how to calculate the frequency or wavelength of electromagnetic waves.

Electromagnetic waves are electric and magnetic disturbances that can be used to transfer energy from a source to an absorber. You use waves from different parts of the **electromagnetic spectrum** in everyday devices and gadgets, including:

- microwave ovens – energy is transferred from a microwave source to the food in the oven, heating it
- radiant heaters – infrared radiation transfers energy from the heater to heat the surroundings.

Electromagnetic waves do not transfer matter. The energy they transfer depends on the wavelength of the waves. This is why waves of different wavelengths have different effects. Figure 1 shows some of the uses of each part of the electromagnetic spectrum.

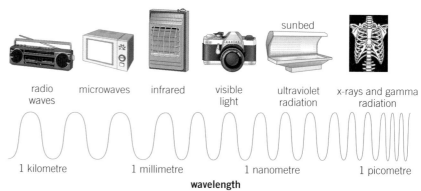

| radio waves | microwaves | infrared | visible light | ultraviolet radiation | x-rays and gamma radiation |

| 1 kilometre | | 1 millimetre | | 1 nanometre | | 1 picometre |

wavelength

(1 nanometre $(1 \times 10^{-9}\,\text{m})$ = 0.000 001 millimetres, 1 picometre = 0.001 nanometres)

Figure 1 *The spectrum is continuous. The frequencies and wavelengths at the boundaries are approximate because the different parts of the spectrum are not precisely defined.*

Waves from different parts of the electromagnetic spectrum have different wavelengths.

- Long-wave radio waves have wavelengths as long as 10 km (10^4 m).
- X-rays and gamma rays have wavelengths as short as a millionth of a millionth of a millimetre (= 0.000 000 000 001 mm or 10^{-15} m).
- Your eyes detect visible light, which is only a limited part of the electromagnetic spectrum (wavelengths of just below 400 nm to above 700 nm).

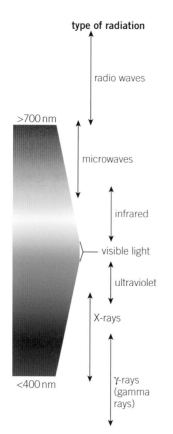

type of radiation

radio waves

>700 nm

microwaves

infrared

visible light

ultraviolet

X-rays

<400 nm γ-rays (gamma rays)

Figure 2 *The electromagnetic spectrum with an expanded view of the visible range*

The speed of electromagnetic waves

All electromagnetic waves travel at a speed of 3.0×10^8 m/s (300 million m/s) through space or in a vacuum. This is the distance the waves travel each second.

You can link the speed of the waves to their frequency and wavelength by using the **wave speed** equation:

Synoptic links

You learnt about the wave speed equation in Topic P11.2.

wave speed, v (m/s) = **frequency, f (Hz)** × **wavelength, λ (m)**
(metres per second, m/s) (hertz, Hz) (metres, m)

You can work out the wavelength λ or the frequency f by rearranging the wave speed equation into:

$$\lambda = \frac{v}{f} \quad \text{or} \quad f = \frac{v}{\lambda}$$

Worked example

A mobile phone gives out electromagnetic waves of frequency 900 million Hz. Calculate the wavelength of these waves.

The speed of electromagnetic waves in air = 300 million m/s.

Solution

Wavelength λ (m) = $\dfrac{\text{wave speed } v \text{ (m/s)}}{\text{frequency } f \text{ (Hz)}}$

$= \dfrac{300\,000\,000 \text{ m/s}}{900\,000\,000 \text{ Hz}} = \textbf{0.33 m}$

Energy and frequency

The wave speed equation shows you that since electromagnetic waves all have a speed of 300 million m/s, the shorter the wavelength of the waves, the higher their frequency. The energy of the waves increases as the frequency increases. So as the wavelength decreases along the electromagnetic spectrum from radio waves to gamma rays, the energy and frequency of the waves increase.

1 a Determine which is greater: the wavelength of radio waves or the wavelength of visible light waves. [1 mark]
   b Describe the speed in a vacuum of different electromagnetic waves. [1 mark]
   c Determine which is greater: the frequency of X-rays or the frequency of infrared radiation. [1 mark]
   d Determine where in the electromagnetic spectrum you would find waves of wavelength 10 millimetres. [1 mark]

2 a Put the following parts of the electromagnetic spectrum in order of increasing frequency:
      infrared   radio   X-rays and gamma rays [1 mark]
   b Determine which parts of the electromagnetic spectrum are missing from the list in **a**. [1 mark]

3 Electromagnetic waves travel through space at a speed of 300 million metres per second. Calculate:
   a the wavelength of radio waves of frequency 600 million Hz
   b the frequency of microwaves of wavelength 0.30 m [4 marks]

4 A distant star explodes and emits visible light and gamma rays simultaneously. Explain why the gamma rays and the visible light waves reach the Earth at the same time. [2 marks]

Go further!

When an atom or a nucleus emits electromagnetic waves, it emits a packet of waves referred to as a *photon*. Einstein imagined a photon like a flying needle and he showed that its energy is proportional to the frequency of the emitted waves. So, the bigger the frequency of the emitted radiation, the greater the energy of each photon.

Study tip

The worked example on this page is an example of where standard form can be useful. The large values for wave speed and frequency could have been written as 3×10^8 m/s and 9×10^8 Hz, respectively. It is worth learning how to do this on your calculator.

Key points

- The electromagnetic spectrum (in order of decreasing wavelength and increasing frequency and energy) is made up of:
  - radio waves
  - microwaves
  - infrared radiation
  - visible light (red to violet)
  - ultraviolet waves
  - X-rays and gamma rays.
- The human eye can only detect visible light. The wavelength of visible light ranges from just below 400 nm to above 700 nm.
- Electromagnetic waves transfer energy from a source to an absorber.
- The wave speed equation $v = f\lambda$ is used to calculate the frequency or wavelength of electromagnetic waves.

P12.2 Light, infrared, microwaves, and radio waves

Learning objectives

After this topic, you should know:

- the nature of white light
- what infrared radiation, microwaves, and radio waves are used for
- what mobile phone radiation is
- why these types of electromagnetic radiation are hazardous.

Light

Light from ordinary lamps and from the Sun is called **white light**. This is because it has all the colours of the visible spectrum in it. The wavelength increases across the spectrum as you go from violet to red. When you look at a rainbow, you see the colours of the spectrum. You can also see them if you use a glass prism to split a beam of white light.

Photographers need to know how shades and colours of light affect the photographs they take.

1 In a film camera, the light is focused by the camera lens on to a light-sensitive film. The film then needs to be developed to see the image of the objects that were photographed.

2 In a digital camera or a mobile phone camera, the light is focused by the lens on to a sensor. This is made up of thousands of tiny light-sensitive cells called pixels. Each pixel gives a dot of the image. The image can be seen on a small screen at the back of the camera. When a photograph is taken, the image is stored electronically on a memory card.

Infrared radiation

All objects emit infrared radiation.

- The hotter an object is, the more infrared radiation it emits.
- Infrared radiation is absorbed by your skin. It can damage, burn, or kill skin cells because it heats up the cells.

Infrared devices

- **Optical fibres** in communications systems usually use infrared radiation instead of visible light. This is because infrared radiation is absorbed less than visible light in the glass fibres.
- Remote control handsets for TV and video equipment transmit signals carried by infrared radiation. When you press a button on the handset, it sends out a sequence of infrared pulses. Infrared radiation is used because suitable infrared pulses can easily be produced and detected electronically.
- Infrared scanners are used in medicine to detect infrared radiation emitted from hot spots on the body surface. These hot areas can mean the tissue underneath is unhealthy.
- You can use infrared cameras to see people and animals in the dark.

  Infrared radiation is used to heat up objects quickly:

  - electric heaters that emit infrared radiation warm rooms quickly
  - electric cookers that have halogen hobs heat up food faster than ordinary hobs because halogen hobs are designed to emit much more infrared radiation than ordinary hobs.

Microwaves

Microwaves have a shorter wavelength than radio waves.

- People use microwaves for communications, for example satellite TV, because they can pass through the atmosphere and reach satellites above the Earth. Microwaves can also carry mobile phone signals.

- Microwave ovens heat food faster than ordinary ovens. This is because microwaves can penetrate into food and are absorbed by the water molecules in the food, heating it. The oven itself does not absorb microwaves as it does not contain any water molecules. It therefore does not become hot like the food it is cooking.

Radio waves

Radio wave frequencies range from about 300 000 Hz to 3000 million Hz (where microwave frequencies start). Radio waves are used to carry radio, TV, and mobile phone signals.

You can also use radio waves instead of cables to connect a computer to other devices such as a printer or a computer mouse. For example, Bluetooth-enabled devices can communicate with each other over a range of about 10 metres without the need for cables.

Microwaves and radio waves can be hazardous because they penetrate people's bodies and can heat the internal parts of the body.

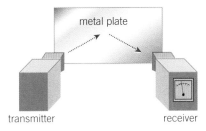

Figure 2 *Detecting microwaves*

1 **a** When you watch a TV programme, name the type of electromagnetic wave that is:
    i detected by the aerial [1 mark]
    ii emitted by the screen. [1 mark]
   b Name the type of electromagnetic wave that is used:
    i to carry signals to and from a satellite [1 mark]
    ii to send signals to a printer from a computer without using a cable. [1 mark]

2 Mobile phones use electromagnetic waves in a wavelength range that includes short-wave radio waves and microwaves.
   a Describe the effect on mobile phone users if remote control handsets operated in this range as well. [1 mark]
   b Explain why the emergency services use radio waves in a wavelength range that no one else is allowed to use. [2 marks]

3 The speed of electromagnetic waves in air is 300 000 km/s. Calculate the wavelength in air of electromagnetic waves of frequency 2400 MHz. [2 marks]

4 Figure 2 shows a microwave receiver being used to detect microwaves from a transmitter. The reading on the receiver meter depends on how much radiation the receiver detects.
   a Describe what the metal plate does to the microwaves. [1 mark]
   b Design a test to find out if microwaves can pass through **i** a metal plate **ii** a thick cardboard sheet. ✿ [4 marks]

P12.3 Communications

Learning objectives

After this topic, you should know:

- why radio waves of different frequencies are used for different purposes
- which waves are used for satellite TV
- how to decide whether or not mobile phones are safe to use
- **❶** What carrier waves are
- how optical fibres are used in communications.

Figure 1 *Sending microwave signals to a satellite*

Figure 2 *A mobile phone mast*

Radio communications

When you use a mobile phone, radio waves carry signals between your mobile phone and the nearest mobile phone mast. The waves used to carry any type of signal are called **carrier waves**. They could be radio waves, microwaves, infrared radiation, or visible light. The type of waves used to carry a signal depends on how much information is in the signal and the distance the signal has to travel. For example, microwaves are used to carry signals via satellites to distant countries.

Radio wavelengths

The radio and microwave spectrum is divided into bands of different wavelength ranges. This is because the shorter the wavelength of the waves:

- the more information they can carry
- the shorter their range (due to increasing absorption by the atmosphere)
- the less they spread out.

Microwaves and radio waves of different wavelengths are used for different communications purposes. Examples include:

- Microwaves are used for satellite phone and TV links, and satellite TV broadcasting. This is because microwaves can travel between satellites in space and the ground. Also, they spread out less than radio waves do, so the signal doesn't weaken as much.
- Radio waves of wavelengths less than about 1 metre are used for TV broadcasting from TV masts because they can carry more information than longer radio waves.
- Radio waves of wavelengths from about 1 metre up to about 100 m are used by local radio stations (and for the emergency services) because their range is limited to the area round the transmitter.
- Radio waves of wavelengths greater than 100 m are used by national and international radio stations because they have a much longer range than shorter-wavelength radio waves.

Mobile phones and electromagnetic radiation

A mobile phone sends out a radio signal when you use it. If the phone is very close to your brain, some scientists think the radiation might affect the brain. Because children have thinner skulls than adults, their brains might be more affected by mobile phone radiation. A UK government report published in May 2000 recommended that the use of mobile phones by children should be limited. More research needs to be conducted to find out if mobile phone users are affected.

More about signals and carrier waves

(Higher)

When you speak, you produce sound waves of different frequencies, so you vary (i.e. modulate) the amplitude and the frequency of the sound waves you produce. In a radio station, a microphone produces an alternating current called an audio signal when sound waves reach it. Figure 3 shows how the signal is transmitted and detected.

- An oscillator supplies carrier waves to the transmitter in the form of an alternating current (a current that repeatedly reverses its direction).
- The audio signal is supplied to the transmitter where it's used to modulate the carrier waves.
- The modulated carrier waves from the transmitter are supplied to the transmitter aerial. The varying alternating current supplied to the aerial causes it to emit radio waves that carry the audio signal.
- When the radio waves are absorbed by a receiver aerial, they induce an alternating current in the receiver aerial, which causes oscillations in the receiver. The frequency of the oscillations is the same as the frequency of the radio waves.
- The receiver circuit separates the audio signal from the carrier waves. The audio signal is then supplied to a loudspeaker, which sends out sound waves similar to the sound waves received by the microphone in the radio station.

Optical fibre communications

Optical fibres are very thin glass fibres. They are used to transmit signals carried by light or infrared radiation. The light rays can't escape from the fibre. When they reach the surface of the fibre, they are reflected back into the fibre (Figure 4).

Compared with radio waves and microwaves:

- optical fibres carry much more information as light has a much shorter wavelength than radio waves, and so can carry more pulses of waves
- optical fibres are more secure because the signals stay in the fibre.

1 **a** Name the types of electromagnetic wave that are used to carry signals along a thin transparent fibre. [1 mark]
   b Explain why signals in an optical fibre are more secure than radio signals. [2 marks]

2 **a** Explain why children could be more affected by mobile phone radiation than adults. [2 marks]
   b i 🄷 Explain what is meant by a carrier wave. [2 marks]
   ii Explain why visible light waves can carry more information than radio waves. [2 marks]

3 Explain why microwaves are used for satellite TV and radio waves are used for terrestrial TV. [3 marks]

4 A local radio station broadcasts at a frequency of 105 MHz.
   a Calculate the wavelength of radio waves of this frequency. The speed of electromagnetic waves in air is 300 000 km/s. [2 marks]
   b Explain why national radio stations broadcast at much lower frequencies. [2 marks]

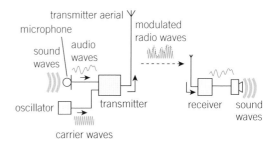

Figure 3 *Using radio waves*

Synoptic links

You learnt about alternating currents in Topic P5.1.

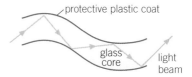

Figure 4 *the reflection of light inside an optical fibre. This is called total internal reflection*

Key points

- Radio waves of different frequencies are used for different purposes because the wavelength (and so the frequency) of waves affects:
  - how far they can travel
  - how much they spread
  - how much information they can carry.
- Microwaves are used for satellite TV signals.
- Further research is needed to evaluate whether or not mobile phones are safe to use.
- 🄷 Carrier waves are waves that are used to carry information. They do this by varying their amplitude.
- Optical fibres are very thin transparent fibres that are used to transmit communication signals by light and infrared radiation.

P12.4 Ultraviolet waves, X-rays, and gamma rays

Ultraviolet waves

Watch your teacher place different-coloured clothes under an ultraviolet lamp. Observe what happens.

- Describe what white clothes look like under a UV lamp.

Safety: The lamp must point downwards so you can't look directly at the glow from it.

Figure 1 *Using an ultraviolet lamp to detect finger prints*

Synoptic links

You learnt about radioactive substances in Topic P7.3 and Topic P7.4.

Study tip

Make sure that you know the dangers, as well as the uses, of the different kinds of electromagnetic waves.

Ultraviolet waves

Ultraviolet (UV) waves lie between violet light and X-rays in the electromagnetic spectrum. Some chemicals emit light as a result of absorbing ultraviolet waves. Posters and ink that glow in ultraviolet light contain these chemicals. Security marker pens containing this kind of ink are used to mark valuable objects. The chemicals absorb ultraviolet waves and then emit visible light.

Ultraviolet waves are harmful to human eyes and can cause blindness. UV wavelengths are smaller than visible light wavelengths. UV waves carry more energy than visible light waves.

Ultraviolet waves are harmful to your skin. For example, too much UV directly from the Sun or from a sunbed can cause sunburn and skin cancer. It can also age the skin prematurely.

- If you stay outdoors in summer, use skin creams to block UV waves and prevent them reaching your skin.
- If you use a sunbed to get a suntan, don't go over the recommended time. You should also wear special goggles to protect your eyes.

X-rays and gamma rays

X-rays and gamma rays both travel straight into substances and can pass through them if the substances are not too dense and not too thick. A thick plate made of lead will stop them.

X-rays and gamma rays have similar properties because they both:

- are at the short-wavelength end of the electromagnetic spectrum
- carry much more energy per second than longer-wavelength electromagnetic waves.

They differ from each other because:

- X-rays are produced when electrons or other particles moving at high speeds are stopped – X-ray tubes are used to produce X-rays
- gamma rays are produced by radioactive substances when unstable nuclei release energy
- gamma rays have shorter wavelengths than X-rays, so they can penetrate substances more than X-rays can.

X-rays are often used to detect internal cracks in metal objects. These kinds of application are usually possible because the more dense a substance is, the more X-rays it absorbs from an X-ray beam passing through it. X-rays are also used in medicine to create images of broken limbs. You will learn more about this in Topic P12.5.

Using gamma rays

High-energy gamma rays have several important uses:

Killing harmful bacteria

1 About 20% of the world's food is lost through spoilage, mostly due to bacteria. Bacteria waste products cause food poisoning. Exposing food to gamma rays kills 99% of disease-carrying organisms, including *Salmonella* (found in poultry) and *Clostridium* (which causes botulism).

2 Exposing surgical instruments in sealed plastic wrappers to gamma rays kills any bacteria on the instruments. This helps to stop infection spreading in hospitals.

Killing cancer cells

Doctors and medical physicists use gamma-ray therapy to destroy cancerous tumours. A narrow beam of gamma rays from a radioactive source (cobalt-60) is directed at the tumour. The beam is aimed at it from different directions to kill the tumour but not the surrounding tissue.

Safety matters

X-rays and gamma rays passing through substances can knock electrons out of atoms in the substance. The atoms become charged because they lose electrons. This process is called ionisation, and so X-rays and gamma rays are examples of ionising radiation.

If ionisation happens to a living cell, it can damage or kill the cell. For this reason, exposure to too many X-rays or gamma rays is dangerous and can cause cancer. High doses kill living cells, and low doses cause gene mutation and cancerous growth.

People who use equipment or substances that produce any form of ionising radiation (e.g., X-rays or gamma rays) must wear a film badge. If the badge shows that it is over-exposed to ionising radiation, its wearer is not allowed to continue working with the equipment for a period of time.

1 a Explain why a crack inside a metal object shows up on an X-ray image. [3 marks]

b Will gamma rays pass through thin plastic wrappers? [1 mark]

c Explain why a film badge used for monitoring radiation needs to have a plastic case, not a metal case. [2 marks]

2 a Explain why ultraviolet waves are harmful. [2 marks]

b i Explain how the Earth's ozone layer helps to protect you from ultraviolet waves from the Sun. [1 mark]

ii Explain why people outdoors in summer need suncream. [3 marks]

3 a Name the types of electromagnetic radiation that can penetrate thin metal sheets. [1 mark]

b Name the metal that can be used most effectively to absorb X-rays and gamma rays. [1 mark]

4 a Explain what is meant by ionisation, and describe one way in which ionisation can occur. 🖉 [3 marks]

b Name the types of electromagnetic radiation that can:

i ionise substances they pass through [1 mark]

ii damage the human eye. [1 mark]

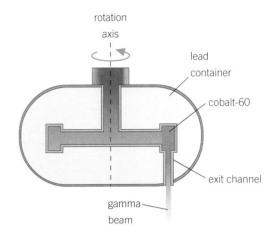

Figure 2 *Gamma treatment – the cobalt-60 source is in a thick lead container. When it is not in use, it is rotated away from the exit channel*

Figure 3 *A film badge tells you how much ionising radiation the wearer has received*

Key points

- Ultraviolet waves have a shorter wavelength than visible light and can harm the skin and the eyes.
- X-rays are used in hospitals to make X-ray images.
- Gamma rays are used to kill harmful bacteria in food, to sterilise surgical equipment, and to kill cancer cells.
- Ionising radiation makes uncharged atoms become charged.
- X-rays and gamma rays damage living tissue when they pass through it.

P12.5 X-rays in medicine

Learning objectives

After this topic, you should know:

- what X-rays are used for in hospitals
- why X-rays are dangerous
- what absorbs X-rays when they pass through the body.

Figure 1 *Taking a chest X-ray*

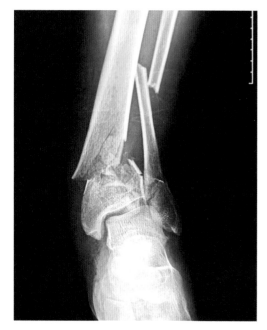

Figure 2 *Spot the break*

Synoptic links

You have learnt about radioactive substances and background radiation in Topics P7.3–7.4

Have you ever broken one of your bones? If you have, you will have gone to your local hospital for an X-ray photograph. X-rays are electromagnetic waves at the short-wavelength end of the electromagnetic spectrum. They are produced in an X-ray tube when fast-moving electrons hit a target. Their wavelengths are about the same as the diameter of an atom.

To make a radiograph or X-ray photograph, X-rays from an X-ray tube are directed at the patient. A lightproof cassette containing a photographic film or a flat-panel detector is placed on the other side of the patient.

- When the X-ray tube is switched on, X-rays from the tube pass through the part of the patient's body under investigation (Figure 1).

- X-rays pass through soft tissue, but they are absorbed by bones, teeth and metal objects that are not too thin. The parts of the film or the detector that the X-rays reach become darker than the other parts. So the bones appear lighter than the surrounding tissue, which appears dark (Figure 2). The radiograph shows a 'negative image' of the bones. A hole or a cavity in a tooth shows up as a dark area in the bright image of the tooth.

- An organ that consists of soft tissue can be filled with a substance called a **contrast medium** that absorbs X-rays easily. This enables the internal surfaces in the organ to be seen on the radiograph. For example, to obtain a radiograph of the stomach, the patient is given a barium meal before the X-ray machine is used (Figure 3). The barium compound is a good absorber of X-rays.

- Lead plates between the tube and the patient stop X-rays reaching other parts of the body. The X-rays reaching the patient pass through a gap between the plates. Lead is used because it is a good absorber of X-rays.

- A flat-panel detector is a small screen that contains a **charge-coupled device (CCD)**. The sensors in the CCD convert X-rays to light. The light rays then create electronic signals in the sensors that are sent to a computer, which displays a digital X-ray image.

Radiation dose

X-rays, gamma rays, and the radiation from radioactive substances all ionise substances they pass through. There are three types of radiation from radioactive substances. You have already learnt about them. Gamma radiation is one of the three types. The other two types are called alpha and beta radiation.

All the different types of ionising radiation are dangerous. The **radiation dose** received by a person is a measure of the damage done to their body by ionising radiation. The radiation dose depends on:

- the type of radiation used
- how long the body is exposed to it
- The energy per second absorbed by the body from the radiation.

For example, alpha radiation inside the body causes ten times more damage than X-rays, when a person is exposed to them for the same length of time.

Radiation dose is measured in sieverts (Sv) or millisieverts (mSv).

High doses of radiation kill living cells. Low doses can cause gene mutation and cancerous growth. There is no evidence of a safe limit below which living cells would not be damaged.

Everyone is exposed to low levels of ionising radiation from background sources such as cosmic radiation from space and radon gas which seeps through the earth from deep underground. Also, workers who use equipment or substances that produce ionising radiation must wear a film badge that tells them how much ionising radiation they have received.

X-ray therapy

Doctors use X-ray therapy to destroy cancerous tumours in the body. Thick plates between the X-ray tube and the body stop X-rays from reaching healthy body tissues. A gap between the plates allows X-rays through to reach the tumour. X-rays for therapy are shorter in wavelength than X-rays used for imaging.

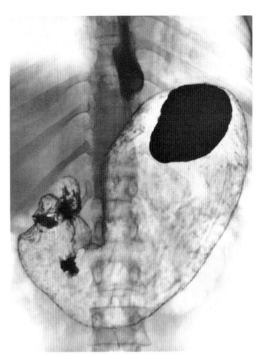

Figure 3 *A coloured X-ray of a stomach ulcer*

Higher

The X-rays used for therapy carry much more energy than X-rays used for imaging. Low-energy X-rays are suitable for imaging because they are absorbed by bones and teeth but they pass through soft tissue and gaps such as cracks in bones. Low-energy X-rays do not carry enough energy to destroy cancerous tumours.

1　a　Explain what a contrast medium is used for when an X-ray photograph of the stomach is taken. [2 marks]
　　b　Describe what X-ray therapy can be used for. [1 mark]

2　When an X-ray photograph is taken, explain why it is necessary:
　　a　to place the patient between the X-ray tube and the film cassette [3 marks]
　　b　to have the film in a lightproof cassette [1 mark]
　　c　to shield the parts of the patient not under investigation from X-rays. Explain what would happen to healthy cells if they were not shielded.  [4 marks]

3　Ⓗ a　Name one way in which X-rays used for X-ray therapy differ from X-rays used for X-ray imaging. [1 mark]
　　　b　Explain why X-rays used for imaging cannot be used for X-ray therapy. [1 mark]

4　The average radiation dose each person receives from ionising radiation is about 2 millisieverts per year. Medical X-rays account for about 13% of this.
　　a　Explain what is meant by radiation dose. [1 mark]
　　b　Estimate the average radiation dose each person receives in one year due to medical X-rays. [1 mark]

Study tip

1 Sv = 1000 mSv.

Study tip

You do not need to remember the unit of radiation dose.

Key points

- X-rays are used in hospitals:
  - to make images of your internal body parts
  - to destroy tumours at or near the body surface.
- X-rays are ionising radiation and so can damage living tissue when they pass through it.
- X-rays are absorbed more by bones and teeth than by soft tissues.

P12 Electromagnetic waves

Summary questions

1 a Place the five types of electromagnetic wave listed below in order of increasing wavelength.

 A Infrared waves

 B Microwaves

 C Radio waves

 D Gamma rays

 E Ultraviolet waves [1 mark]

 b Name the type/s of electromagnetic radiation listed in **a** that:

 i can be used to send signals to and from a satellite [1 mark]

 ii ionise substances when they pass through them [1 mark]

 iii are used to carry signals in thin transparent fibres. [1 mark]

2 a The radio waves from a local radio station have a wavelength of 2.9 m in air. The speed of electromagnetic waves in air is 300 000 km/s.

 i Write the equation that links frequency, wavelength, and wave speed. [1 mark]

 ii Calculate the frequency of the radio waves. [2 marks]

 b A certain local radio station transmitter has a range of 30 km. Describe and explain the effect on the range if the power supplied to the transmitter is reduced. [3 marks]

3 The signal from a TV remote control is transmitted as infrared radiation of a certain wavelength. The signal to and from a mobile phone is transmitted as radio waves of a certain wavelength.

 a **i** Write down two differences between infrared radiation and radio waves. [2 marks]

 ii Write down one similarity between infrared radiation and radio waves. [1 mark]

 b Infrared lamps are used in a car factory to dry the paint on metal car panels after they have been sprayed with paint.

 i Explain the property of infrared radiation that makes it suitable for this purpose. [1 mark]

 ii Give one reason why microwaves would not be suitable for this purpose. [2 marks]

 c Signals can be transmitted along optical fibres using infrared radiation or visible light. Explain why a signal transmitted along an optical fibre becomes weaker the further it travels along the fibre. [3 marks]

4 The figure shows an X-ray source that is used to direct X-rays at a broken leg. A photographic film in a lightproof wrapper is placed under the leg. When the film is developed, an image of the broken bone is observed.

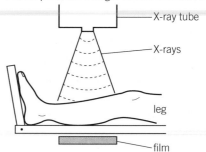

 a **i** Explain why an image of the bone is seen on the film. [3 marks]

 ii Why is it possible to see the fracture on the image. [1 mark]

 b When an X-ray photograph of the stomach is taken, the patient is given food containing barium before the photograph is taken.

 i Explain why it is necessary for the patient to be given this food before the photograph is taken. [3 marks]

 ii The exposure time for a stomach X-ray must be shorter than the X-ray time for a limb. Explain why. [2 marks]

 iii Low-energy X-rays from the X-ray tube can be absorbed by placing a metal plate between the patient and the X-ray tube. Such X-rays would otherwise be absorbed by the body. Describe the benefit of removing such low-energy X-rays in this way. [2 marks]

5 a What is meant by ionisation? [1 mark]

 b Name the **two** types of electromagnetic radiation that can ionise substances. [1 mark]

 c Give **two** reasons why ionising radiation is harmful. [2 marks]

6 a Some chemicals can emit light as a result of absorbing ultraviolet waves. Describe and explain how these chemicals could be used as invisible ink. In your explanation, including the type of radiation that is absorbed and the type of radiation that is emitted. [3 marks]

 b Explain why a beam of infrared radiation cannot be used to carry signals to a detector that is more than a few metres from a transmitter. [2 marks]

Practice questions

01 **Figure 1** shows the electromagnetic spectrum.

radio waves	microwaves		light	ultraviolet		gamma

01.1 Give the names of the two missing waves. [2 marks]

01.2 Complete the sentences using words from the box below. Each word can be used once, more than once, or not at all.

lower	the same	longer	sound
shorter	energy	higher	faster

Radio waves have a _____ wavelength and _____ frequency than other electromagnetic waves.

[2 marks]

Radio waves have _____ speed in air compared with microwaves. [1 mark]

Radio waves transfer _____ from place to place.

[1 mark]

01.3 Electromagnetic waves are used in many applications. Match the correct wave to one use of the wave.

microwave		kills cancer cells
ultraviolet		used in mobile phones
gamma		used in sunbeds

[2 marks]

02 The following headline appeared in a local newspaper.
Councillor Jones says that all mobile phone masts near local schools should be banned. They produce harmful radiation.

02.1 Satellites are used to send messages around the world using microwaves. Give one reason why microwaves are used but not radio waves. [1 mark]

02.2 A spokesperson from a mobile phone company wrote to the newspaper stating that there are no risks from the masts. Suggest one reason why this statement should be treated with caution. [1 mark]

02.3 A study asked 42 000 people if they used a mobile phone and whether they had cancer. The conclusion was that there was no risk of cancer by using mobile phones. Evaluate the method used to determine whether the conclusion is valid and suggest possible improvements. [4 marks]

02.4 Microwave radiation is classed as non-ionizing radiation. Explain the dangers of ionizing radiation. [2 marks]

03 Gamma radiation and X-rays are used in medicine.

03.1 Surgical instruments are sealed in a plastic bag and then irradiated with gamma rays. Describe how this method keeps the instruments sterile. [2 marks]

03.2 Gamma knife surgery uses many low intensity beams of gamma radiation that come from different directions. All of the beams are focussed onto a tumour in the patient's head. Describe how this procedure targets the tumour but reduces the risk to the patient. [3 marks]

Figure 1

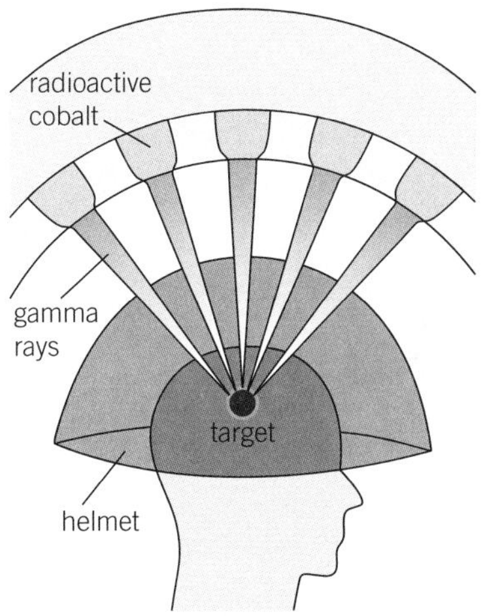

03.3 Describe the properties of X-rays that make them suitable for detecting a broken arm. [2 marks]

04 Sunglasses manufacturers predict that within 10 years, everyone will wear UV protection sunglasses when outside.

04.1 Which one of the following statements is the most likely reason for the prediction made by the manufacturers?

Opticians and other experts will make people aware of the dangers.
The style of sunglasses will be very modern.
The level of sunlight in summer will increase.

[1 mark]

04.2 Exposure to too much UV radiation is known to increase the risk of skin cancer. Suggest what precautions a golfer and snow skier should take when playing their sports. [2 marks]

04.3 Calculate the wave speed of a beam of ultraviolet radiation.
The frequency of the wave is 8×10^{14} Hz and the wavelength is 3.75×10^{-7} m.

Learning objectives

After this topic, you should know:

- the force rule for two magnetic poles near each other
- the pattern of magnetic field lines around a bar magnet
- what induced magnetism is
- why steel, not iron, is used to make permanent magnets.

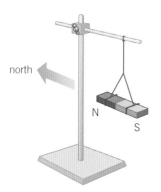

Figure 1 *Checking the poles of a bar magnet*

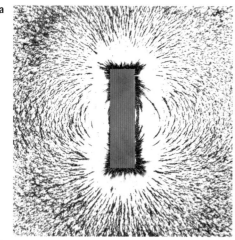

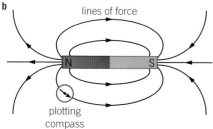

Figure 2 *The magnetic field of a bar magnet a Using iron filings b Using a plotting compass*

About magnets

A magnetic compass is a tiny magnetic needle pivoted at its centre. Because of the Earth's magnetic field, one end of the compass always points north, and the other end always points south. The end of a plotting compass or a bar magnet (Figure 1) that points north is the 'north-seeking' pole, usually called its north pole (N-pole), and the other end is the 'south-seeking' pole, its south pole (S-pole).

Investigating bar magnets

1 Suspend a bar magnet as shown in Figure 1 and label the end that points north as its N-pole.

2 Hold the N-pole of a second bar magnet near the suspended bar magnet. You should find it attracts the S-pole of the suspended bar magnet and repels the N-pole.

3 Repeat the above test using the S-pole of the second bar magnet. You should find it attracts the N-pole of the suspended bar magnet and repels its S-pole.

The tests above show the general rule that:

Like poles repel. Unlike poles attract.

Magnetic materials

Any iron or steel object can be magnetised (or demagnetised if it's already magnetised). Only a few other materials (for example cobalt and nickel) can be magnetised and demagnetised. Permanent magnets are made of steel because magnetised steel does not lose its magnetism easily.

Magnetic fields

If a sheet of paper is placed over a bar magnet and iron filings are sprinkled onto the paper, the filings form a pattern of lines. The region around the magnet is called a **magnetic field**. Any other magnetic material placed in this space experiences a force caused by the first magnet.

In Figure 2:

- the iron filings form lines as shown in Figure 2a that end at or near the poles of the magnet. These lines are **magnetic field lines**, also called lines of force. The lines are more concentrated at the poles than elsewhere. This is because the field is strongest at the poles.

- a plotting compass placed in the magnetic field aligns itself along a magnetic field line, pointing in a direction away from the N-pole of the magnet and towards the magnet's S-pole, as shown in Figure 2b. For this reason, the direction of a line of force is always from the north pole of the magnet to its south pole.

The further the plotting compass is from the magnet, the less effect the magnet has on the plotting compass. This is because the greater the distance from the magnet, the weaker the strength of the magnetic field.

Induced magnetism

An unmagnetised magnetic material can be magnetised by placing it in a magnetic field. The magnetic field is said to **induce magnetism** in the material. For example, an unmagnetised iron rod placed in line with a bar magnet becomes a magnet with poles at each end. The nearest poles of the rod and the bar magnet always have opposite polarity.

Induced magnetism will cause a force of attraction between any unmagnetised magnetic material placed near one end of a bar magnet. The force is always an attractive force whichever end of the bar magnet is nearest to the material.

1 A bar magnet XY is freely suspended in a horizontal position so that end X points north and end Y points south.
   a Give the magnetic polarity of:
      i end X   [1 mark]   ii end Y.   [1 mark]
   b End P of a second bar magnet PQ placed near end X of bar magnet XY repels end X and attracts end Y. Give the magnetic polarity of end P and give a reason for this observation.   [2 marks]
2 The tip of an iron nail is held in turn near each end of a plotting compass needle. Write whether the tip of the nail is a N-pole, a S-pole, or is unmagnetised in each of the following possible observations:
   a i the N-pole of the compass needle is repelled by the tip of the nail and the S-pole is attracted by it
      ii the N-pole of the compass needle is attracted by the tip of the nail and the S-pole is repelled by it
      iii the N-pole of the plotting compass is attracted by the tip of the nail and the S-pole is also attracted by it.   (3 marks)
   b Explain why the tip of the nail in **a iii** attracts both poles of the plotting compass.   [2 marks]
3 Figure 4 shows a bar magnet XY and a plotting compass near end Y of the bar magnet. The needle of the plotting compass points towards end Y of the bar magnet.

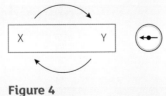

Figure 4

   a Give the magnetic polarity of each end of the bar magnet.   [1 mark]
   b If the bar magnet was rotated gradually about its centre through 180°, describe and explain the effect on the direction of the plotting compass needle.   [3 marks]
4 Design an experiment using a plotting compass and a ruler to find out which of two bar magnets is stronger. Draw a diagram to aid your explanation. 🖊   [4 marks]

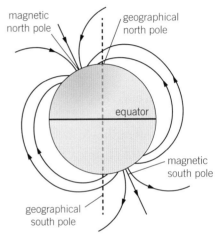

Figure 3 *The Earth's magnetic field. Scientists have plotted the Earth's magnetic field accurately. The pattern is like that of a bar magnet. But the Earth is partly molten inside, and this helps to explain why the magnetic poles move about gradually*

Plotting a magnetic field
Mark a dot near the north pole of a bar magnet. Place the tail of the compass needle above the dot, and mark a second dot at the tip of the needle. Repeat the procedure with the tail over the new dot each time until the compass reaches the S-pole of the magnet. Draw a line through the dots and mark the direction from the N-pole to the S-pole. Repeat the procedure for further lines.

Key points

- Like poles repel, and unlike poles attract.
- The magnetic field lines of a bar magnet curve around from the north pole of the bar magnet to the south pole.
- Induced magnetism is magnetism created in an unmagnetised magnetic material when the material is placed in the magnetic field.
- Steel is used instead of iron to make permanent magnets because steel does not lose its magnetism easily, but iron does.

P13.2 Magnetic fields of electric currents

Learning objectives

After this topic, you should know:

- the pattern of the magnetic field around a straight wire carrying a current, and in and around a solenoid
- how the strength and direction of the field varies with position and with the current
- what a uniform magnetic field is
- what an electromagnet is.

Fields around a current carrying wire

Use the arrangement shown in Figure 1 to observe the effect of:

1 **Reversing the current**
You should find that the plotting compass reverses its direction. This shows that the magnetic field lines reverse direction when the direction of the current is reversed.

2 **Moving the plotting compass away from the wire**
You should find that it points more towards 'magnetic north'. The Earth's magnetic field has a bigger effect the further the compass is from the wire. This is because the strength of the magnetic field of the wire decreases as the distance from the wire increases.

3 **Increasing the current**
To vary the current, connect a variable resistor in series with the battery and the wire. As you increase the current you should find the magnetic field becomes stronger everywhere. You can tell this if you increase the current gradually from zero, the plotting compass turns more and more away from the North as the current becomes stronger. This is because the field has a bigger effect on the plotting compass than the Earth's magnetic field.

Safety: Make sure the wire does not get too hot.

The magnetic field near a current-carrying wire

When an electric current passes along a wire, a magnetic field is set up around the wire. Figure 1 shows how you can find the pattern of the magnetic field around a long straight wire by using a plotting compass. The lines of force caused by a straight current-carrying wire are a series of concentric circles. These circles are centred on the wire in a plane that is perpendicular to the wire.

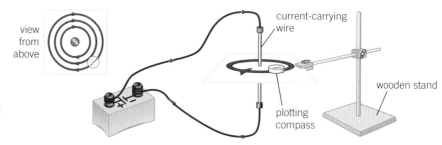

Figure 1 *The magnetic field near a long straight wire. To eliminate magnetism caused by nearby iron objects, use a wooden stand (or any other non-ferrous object) to support the cardboard sheet so that it's horizontal*

You can use the corkscrew rule shown in Figure 2 to remember the direction of the magnetic field for each direction of the current. Reversing the direction of the current reverses the direction of the magnetic field.

Figure 2 *The corkscrew rule*

Solenoids

A **solenoid** is a long coil of insulated wire (Figure 3). Solenoids are used in lots of devices where a strong magnetic field needs to be produced. The magnetic field is produced in and around the solenoid when a current is passed through the wire. The magnetic field:

- increases in strength if the current is increased
- reverses its direction if the current is reversed.

Inside the solenoid

The magnetic field is much stronger than if the wire was straight. The field lines are parallel to the axis of the solenoid, and they are all in the same direction (i.e., **uniform**). The magnetic field inside a solenoid is strong and uniform.

Outside the solenoid

The magnetic field lines bend around from one end of the solenoid to the other end of the solenoid. The magnetic field outside is like the field of a bar magnet, except that each field line is a complete loop because it passes through the inside of the solenoid.

Figure 3 shows how you can find the polarity of each end of the solenoid from the direction of the current.

Electromagnets

An **electromagnet** is a solenoid in which the insulated wire is wrapped around an iron bar (the core). When a current is passed along a wire, a magnetic field is created around the wire. Because of this, the magnetic field of the wire magnetises the iron bar. When the current is switched off, the iron bar loses most of its magnetism.

1 **a** Sketch the pattern of the magnetic field lines near a vertical wire carrying current upwards [2 marks]
   b Show on your sketch a plotting compass near the wire and indicate the direction which it points when there is a current in the wire. [1 mark]
2 **a** Describe how an insulated wire and an iron bar can be used to produce a strong magnetic field when a current is passed through the wire. [2 marks]
   b Explain why iron, and not steel, is used for the core of an electromagnet. [2 marks]
3 A plotting compass is placed near a straight wire carrying a current as shown in Figure 1.
   a Describe the effect on the plotting compass of reversing the current in the wire. [1 mark]
   b Describe and explain how the direction in which the plotting compass points would change if it was gradually moved away from the wire. [3 marks]
4 **a** Sketch the pattern of the magnetic field lines in and around a solenoid when there is a current in the solenoid. [3 marks]
   b i Show on your sketch a plotting compass at one end of the solenoid and indicate the direction that the plotting compass points. [2 marks]
   ii Describe and explain the effect on the plotting compass of reducing the current in the solenoid. [3 marks]

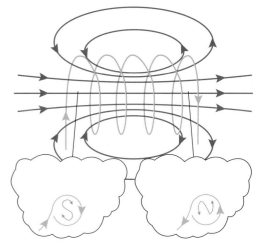

Figure 3 *The magnetic field of a solenoid. Looking at each end, the S-pole is the end where the current is clockwise, and the N-pole is the end at which the current is anticlockwise.*

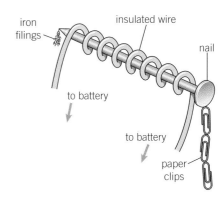

Figure 4 *A simple electromagnet*

Key points

- The magnetic field lines *around a wire* are circles centred on the wire in a plane perpendicular to the wire.
- The magnetic field lines *in a solenoid* are parallel to its axis and are all in the same direction. A uniform magnetic field is one in which the magnetic field lines are parallel.
- Increasing the current makes the magnetic field stronger. Reversing the direction of the current reverses the magnetic field lines.
- An electromagnet is a solenoid that has an iron core. It consists of an insulated wire wrapped around an iron bar.

P13.3 The motor effect

Learning objectives

After this topic, you should know:

- how to change the size and reverse the direction of the force on a current-carrying wire in a magnetic field
- how a simple electric motor works
- what is meant by magnetic flux density
- how to calculate the force on a current-carrying wire.

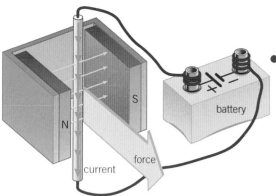

Figure 1 *An experimental setup you could use to investigate the motor effect*

Study tip

Make sure you can use Fleming's left-hand rule to decide the direction of the force on a current-carrying conductor.

You use electric motors lots of times every day. Using a hairdryer, an electric shaver, a refrigerator pump, and a computer hard drive are just a few examples. All these electrical appliances contain an electric motor. The electric motor works because a force can act on a wire (or any other conductor) in a magnetic field when a current is passed through the wire. This is called the **motor effect**.

- The size of the force can be increased by:
  - increasing the current
  - using a stronger magnet.
- The size of the force depends on the angle between the wire and the magnetic field lines. The force is:
  - greatest when the wire is perpendicular to the magnetic field
  - zero when the wire is parallel to the magnetic field lines.
- The direction of the force is always at right angles to the wire and the field lines. Also, the direction of the force is reversed if either the direction of the current or the magnetic field is reversed. Figure 2 shows **Fleming's left-hand rule**, which tells you how these directions are related to each other.

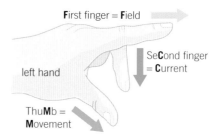

Figure 2 *Fleming's left-hand rule. Hold your fingers at right angles to each other. You can use this rule to work out the direction of the force (i.e. the movement) on the wire.*

Magnetic flux density

The **magnetic flux density** of a magnetic field is the measure of the strength of the magnetic field. The symbol B is used for magnetic flux density and the unit is the tesla (T).

In Figure 1, the size of the force F on the conductor depends on:

- the current I in the conductor
- the length l of the conductor
- the magnetic flux density B of the magnetic field.

You can write the following equation for the force:

$$\text{force, } F = \text{magnetic flux density, } B \times \text{current, } I \times \text{length, } l$$

(newtons, N)   (tesla, T)   (amperes, A)   (metres, m)

Worked example

In Figure 1, the magnetic flux density is 0.032 T, the length of the conductor in the field is 40 mm, and the current through the conductor is 2.2 A. Calculate the force on the conductor.

Solution

The length in metres of the conductor is 0.040 m.

Using $F = B I l$ gives:

$$F = 0.032\,T \times 2.2\,A \times 0.040\,m$$

$$= 0.028\,N$$

The electric motor

An electric motor is designed to use the motor effect. You can control the speed of an electric motor by changing the current. Also, you can reverse the direction the motor turns in by reversing the current.

The simple motor in Figure 3 has a rectangular coil of insulated wire (the armature coil) that is forced to rotate. The coil is connected to the battery by two metal or graphite brushes. The brushes press onto a metal **split-ring commutator** fixed to the coil. Graphite is a form of carbon that conducts electricity and is very slippery. Graphite causes very little friction when it is in contact with the rotating commutator.

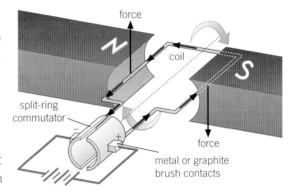

Figure 3 *The electric motor*

When a current is passed through the coil, the coil spins because:

● a force acts on each side of the coil due to the motor effect

● the force on one side is in the opposite direction to the force on the other side.

The split-ring commutator reverses the current around the coil every half-turn of the coil. Because the sides swap over each half-turn, the coil is pushed in the same direction every half-turn.

1 Explain why the coil of a simple electric motor rotates continuously when the motor is connected to a battery. [3 marks]

2 **a** Explain why a simple electric motor connected to a battery reverses if the battery connections are reversed. [2 marks]

   b Determine whether or not an electric motor would run faster if the coil was wound on:

   i a plastic block [1 mark]

   ii an iron block, instead of a wooden block. [1 mark]

3 A force is exerted on a straight wire when a current is passed through it and it is at right angles to the lines of a magnetic field. Describe how the force changes if the wire is turned through 90° until it is parallel to the field lines. [3 marks]

4 A straight wire is placed in the magnetic field of a U-shaped magnet as shown in Figure 1. The length of the wire in the field is 35 mm. When a current of 1.8 A is in the wire, a force of 0.024 N acts on the wire due to the field. Calculate the magnetic flux density of the field. [2 marks]

Key points

● In the motor effect, the force is:
  ▪ increased if the current or the strength of the magnetic field or the length of the conductor is increased
  ▪ reversed if the direction of the current or the magnetic field is reversed

● An electric motor has a coil that turns when a current is passed through it.

● Magnetic flux density is a measure of the strength of a magnetic field.

● To calculate the force on a current-carrying conductor at right angles to the lines of a magnetic field, use the equation $F = B I l$.

Summary questions

1 Two identical bar magnets are placed end-to-end on a sheet of paper on a table with a gap between them with unlike poles facing each other.

   a Draw the arrangement and the pattern of the magnetic field lines in the gap. [1 mark]

   b A plotting compass is placed in the gap at equal distance from the two magnets at a short distance from the midpoint of the gap. On your drawing, show the plotting compass in this position and show the direction in which it points. [1 mark]

2 **a** 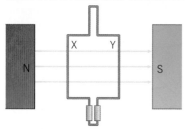 Figure 1 shows a rectangular coil of wire in a magnetic field viewed from above. When a direct current passes clockwise around the coil, a downward force acts on side **X** of the coil.

Figure 1

   i Write the direction of the force on side **Y** of the coil. [1 mark]

   ii Describe the force on each side of the coil parallel to the magnetic field lines. [1 mark]

   iii Describe the effect of the forces on the coil. [1 mark]

3 **a** A vertical wire is in a horizontal magnetic field of magnetic flux density 55 mT. The length of the wire in the field is 45 mm. Calculate the force on the wire when a current of 6.5 A is in the wire. [2 marks]

   b Describe how the force on the wire would change if the current had been:

   i reduced to 3.0 A without changing its direction? [2 marks]

   ii reversed and adjusted to 3.0 A? [1 mark]

4 **a** Cables at a potential difference of 100 000 V are used to transfer 1 000 000 W of electrical power in a grid system.

   i Calculate the current in the cable. [2 marks]

   ii If the potential difference had been 10 000 V, calculate how much current would be needed to transfer the same amount of power. [1 mark]

   b **H** Explain why power is transmitted through the National Grid at a high potential difference rather than a low potential difference. [3 marks]

5 **a** The Earth's magnetic field lines are concentrated at the two magnetic poles and loop round above the surface from one pole to the other pole.

   i Name the element inside the Earth that causes the Earth's magnetic field. [1 mark]

   ii Give the direction of the Earth's magnetic field at the magnetic north pole. [1 mark]

   iii The magnetic poles are known to move slowly. Explain what this tells you about the physical state of the element inside the Earth that causes the Earth's magnetism? [1 mark]

   b The magnetic flux density of the Earth at a certain location on the Earth's surface is 85 μT. A 100 m cable at this location is aligned at right angles to the Earth's magnetic field. Calculate the force on the cable when there is a current of 40 A in it. [2 marks]

6 **a** Figure 2 shows an overhead view of a plotting compass near a vertical wire. When a steady current is in the wire, the plotting compass points in the direction shown.

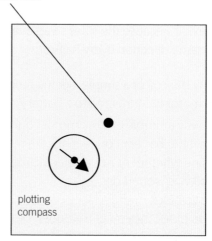

Figure 2

   i Copy the diagram, draw two magnetic field lines near the wire, and indicate the direction of the current. [3 marks]

   ii Describe and explain what would happen to the plotting compass if the current in the wire was reversed. [2 marks]

Practice questions

01.1 **Figure 1** shows five situations A, B, C, D, and E where bars of different materials are placed next to each other. Some of the bars are magnets and some are non-magnets.

Figure 1

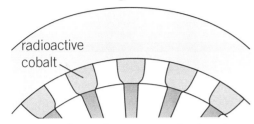

radioactive cobalt

Determine for each situation whether the two bars are repelled, attracted or nothing happens. [5 marks]

01.2 Complete the sentences using the correct words from the box.

increases	an induced	loses
a permanent	charges	changes

An induced magnet _____ its magnetism when moved away from a permanent magnet.
A soft iron bar becomes _____ magnet when placed next to another magnetic material.
[2 marks]

01.3 A small bar magnet is suspended by thin string so that it is free to spin. The south pole of the magnet is marked. When the magnet comes to rest, describe the direction the ends of the magnet will point to and explain why. [2 marks]

02 A U-shaped magnet, which has a 12 mm section of a copper wire between its poles is in series with a battery and a switch.

Figure 2

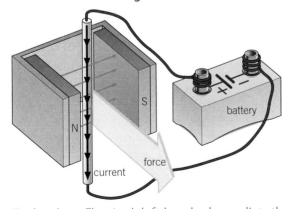

02.1 Explain how Fleming's left-hand rule predicts the direction of the force when the switch is closed.
[3 marks]

02.2 The magnetic flux density between the poles of the magnet is 0.100 T. Calculate the force on the wire when the current in it is 5.2 A [2 marks]

03 A teacher demonstrated how changing the current in a solenoid affects the magnetic force around the solenoid. He used the equipment in **Figure 3**.

Figure 3

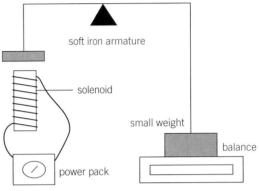

soft iron armature

solenoid

small weight

balance

power pack

03.1 Describe how when a current flows through the solenoid the small weight is raised off the balance.
[3 marks]

03.2 The teacher changed the current through the solenoid and recorded the reading on the balance. Draw a graph of the results. [3 marks]

Current in A	0	2	4	6	8	10
Weight on balance in N	5	2.5	1.2	0.8	0.5	0.0

03.3 What was the weight of the small weight on the balance. [1 mark]

03.4 Name three factors that were kept constant.
[3 marks]

03.5 Name the independent variable. [1 mark]

03.6 Name the dependent variable. [1 mark]

03.7 Suggest a relationship between the current through the solenoid and the magnetic field produced. [2 marks]

04 ⓗ The coil of an electric motor is held between the poles of a magnet and is connected to a switch and a 9 volt battery. When the switch is closed, the left-hand side of the coil experiences an upward force that makes the coil rotate in a clockwise direction.

04.1 Give two ways of making the coil rotate in an anti-clockwise direction [2 marks]

04.2 Give two ways of increasing the force on the coil in the electric motor. [2 marks]

Paper 1 questions

01 A student read that all hot water tanks should be insulated to reduce the rate of energy transfer to the surroundings. She investigated the rate of energy transfer from a beaker of hot water using a series of tests. For each test, she used an insulation material of different thickness.

Figure 1

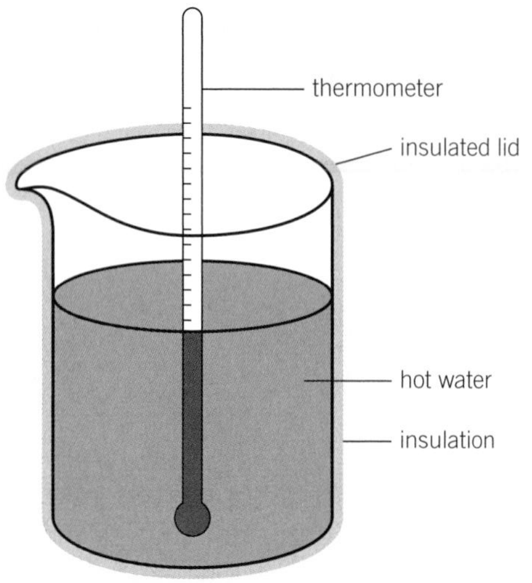

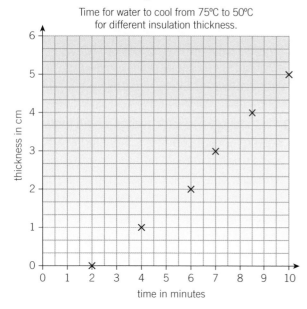

01.1 Give the independent and dependent variables in the investigation. [2 marks]

01.2 Suggest **two** things that should be kept constant. [2 marks]

01.3 Determine what resolution thermometer the student should choose. [1 mark]

01.4 The student plotted her results on graph paper. Copy the graph below and draw the line of best fit. [1 mark]

Figure 2

Time for water to cool from 75°C to 50°C for different insulation thickness.

01.5 Determine the time for the water to cool from 75°C to 50°C without insulation. [1 mark]

01.6 Suggest the relationship between the thickness of insulation and the rate of energy transfer from a hot water tank. [2 marks]

01.7 Suggest **one** improvement that could be made to the investigation. [1 mark]

02 A warm plate stacker is used in a hotel. The two identical springs extend as plates are added, and they retract as plates are removed from the stack. The plates are kept warm by the electric heater at the bottom.

Figure 3

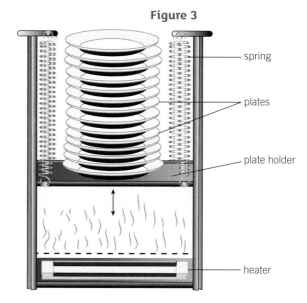

02.1 Calculate the energy stored in the elastic potential energy store of the springs when each spring extends by 0.35 m. The spring constant of each spring is 225 N/m. [2 marks]

02.2 The heater is connected to a 230 V supply, and has 2.17 A of current flowing through it. Calculate the power of the heater. [2 marks]

02.3 Calculate the resistance of the resistance wire used in the heater. [2 marks]

02.4 Suggest **one** reason why it is important to control the temperature of the heater. [1 mark]

02.5 Calculate the temperature of 15 kg of warm plates after 294 000 joules of energy are supplied to the heater. The plates start at a temperature of 15 °C. The specific heat capacity of the plate material is 980 J/kg°C. [3 marks]

03.1 The graph shows how the number of nuclei in a sample of the radioactive isotope plutonium-238 changes with time.

Figure 4

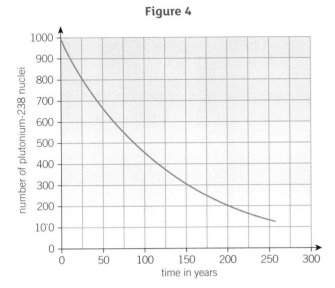

Copy and use the graph to find the half-life of plutonium-238.
Show clearly on the graph how you obtain your answer. [2 marks]

03.2 Plutonium-238 decays by emitting alpha particles. What is an alpha particle? [1 mark]

03.3 Plutonium-238 is highly dangerous. A tiny amount taken into the body is enough to kill a human. Plutonium-238 is unlikely to cause any harm if it is outside the body but is likely to kill if it is inside the body.
Explain why. [3 marks]
AQA, 2009

04 Some elements have radioactive isotopes.

04.1 Describe the similarities and differences between the structure of a radioactive isotope with the structure of a stable isotope of the same element. [3 marks]

04.2 Copy and complete the diagram below to show the radioactive decay of plutonium-238. [3 marks]

$$^{238}_{X}\text{Plutonium} \rightarrow ^{Y}_{92}\text{Uranium} + ^{4}_{2}\text{Z}$$

05 A hill farmer has used a mountain stream to build a system that generates electricity.

05.1 Give the name of the type of energy resource the farmer is using. Choose one word from the box. [1 mark]

Geothermal	Water	Hydroelectric	Uphill

05.2 Evaluate the different methods of generating electricity during the day. You should include environmental issues, reliability and costs. [6 marks]

	Coal	Gas	Nuclear	Wind	Tidal
Cost (£/MWh)	100–155	60–130	80–105	150–210	155–390
Start up Time (hours)	18 hours	3 hours	70 hours	variable	12 hours

06 Energy can be stored or transferred.

06.1 Complete the sentences below using words from the box.

> gravitational potential   kinetic   thermal
> chemical   elastic potential

An aeroplane's _____ energy store increases as it gains height.

A train's _____ energy store increases as it accelerates.

The _____ energy store of the brakes in a car increases as the car slows down.

A gas boiler uses energy from its _____ energy store to boil water. [4 marks]

06.2 A child sits on a swing and is given a push.

Figure 5

Describe the energy transfers and changes to energy stores as the swing moves from right to left [5 marks]

06.3 Suggest **one** reason why the swing eventually comes to a stop [1 mark]

07 A group of students build a bird feeder of length 0.2 m and mass 1.3 kg and attach it to a rope. The rope is used to pull the feeder up a tree and then lower the feeder to refill it with seeds.

Figure 6

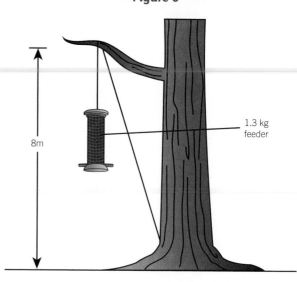

1.3 kg feeder

8m

07.1 Calculate the weight of the bird feeder. [2 marks] Gravitational field strength on Earth is 9.8 N/kg.

07.2 Calculate the increase of the gravitational potential energy store of the bird feeder when it is pulled up to the branch. [2 marks]

07.3 **H** A student lets go of the rope as the bird feeder reaches the branch. Calculate the velocity of the bird feeder as it hits the ground. [3 marks]

07.4 The table shows the frequency of refilling the feeder when the feeder is suspended at different heights.

Refills per week	2	3	5	5	2
Height of feeder	2 m	4 m	6 m	7 m	8 m

Draw conclusions about the relationship between the height of the feeder and the number of refills needed. [3 marks]

08.0 A gardener replaces the cable on a plastic electric lawnmower. The new cable from the lawnmower is connected to 3-pin plug.

Figure 7

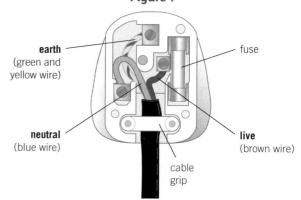

earth (green and yellow wire)

fuse

neutral (blue wire)

live (brown wire)

cable grip

08.1 Comment on whether the gardener needed to use a three-core cable to connect the 3-pin plug to the lawnmower. [1 mark]

08.2 Describe the functions of the three wires connected to the plug. [3 marks]

08.3 Calculate the potential difference between each wire and Earth. [3 marks]

08.4 Explain how the gardener may get an electric shock if the cable is cut whilst she is using the lawnmower. [3 marks]

09.0 Athletes can suffer from tendon and ligament injuries. Recent scientific advances include injecting bone marrow stem cells into the injury to promote healing. A radioactive tracer is inserted into the stem cells to quantify the number of cells retained in the tendon and the distribution of stem cells around the body.

> Technetium-99 has a half-life of 6 hours and emits gamma radiation.
> Radon-224 has a half-life of 3.6 days and emits alpha radiation.

09.1 Which radioactive tracer would you choose? Explain your answer. [3 marks]

09.2 🄷 A sample of cobalt-60 is left to decay for 27 days. Calculate what fraction of the original sample will remain.
Cobalt-60 has a half life of 5.27 days. [2 marks]

09.3 When radioactive materials are used in industry or medicine, technicians must be aware that radioactive contamination can occur.
Describe what is meant by radioactive contamination. [2 marks]

09.4 Suggest **one** reason why it is important for scientists to publish the results of any research on cancer treatment. [1 mark]

10.0 An irregular shaped ornament is sold as solid gold.

Figure 8

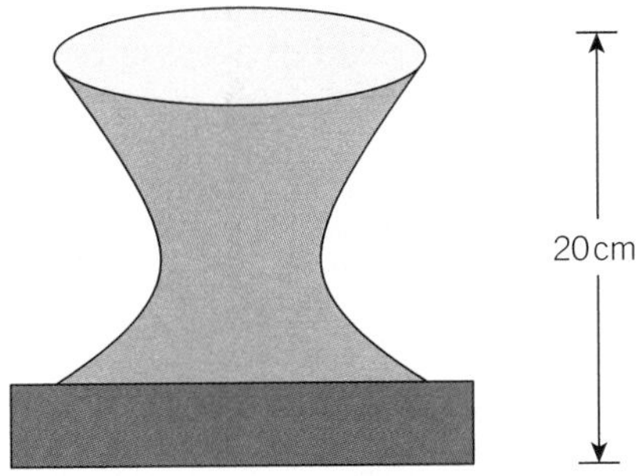

20 cm

10.1 Describe how the volume of the ornament could be determined [2 marks]

10.2 Write down the equation that links density, mass, and volume. [1 mark]

10.3 A similar ornament was found to have a volume of 720 cm³. The mass of the object was 13.91 kg. Decide whether the ornament is made of solid gold. The density of gold is 19.32 g/cm³. Give a reason for your answer. [3 marks]

11.0 A waiter tips a bag of ice cubes into a large bowl in a warm room.

11.1 Describe in terms of molecules how the ice cubes change into water. [4 marks]

11.2 Calculate the latent heat of fusion of ice when 2.45 kg of ice are changed into water. The energy supplied is 8.183×10^6 joules. [3 marks]

12.0 The three-pin plug has been removed from the microwave oven in Figure 9.

Figure 9

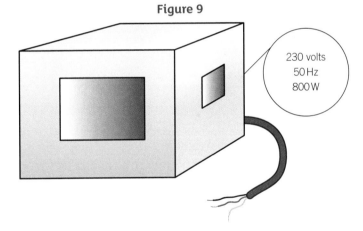

230 volts
50 Hz
800 W

12.1 Give the names of the three coloured wires seen at the end of the three-core cable. [3 marks]

12.2 Describe the purpose of each of the three wires in the three-core cable. [3 marks]

12.3 The microwave oven has a rating plate attached to the side. It gives a value of 50 Hz.
Explain what this value means. [2 marks]

12.4 Calculate the energy transferred by the microwave oven in five minutes. [3 marks]

12.5 Calculate the charge through the microwave oven during the five minutes. You answer should include the units. [4 marks]

Paper 2 questions

01 Electromagnetic waves form a continuous spectrum.

Radio waves	A	Infrared waves	Visible light	Ultraviolet	X-rays	B

01.1 Gives the names of the two types of waves, **A** and **B**, that are missing from the table.　[2 marks]

01.2 Choose words from the box to complete the sentences.　[4 marks]

light　water　air　a vacuum　transverse
the same　energy　longitudinal

All electromagnetic waves transfer _____.
All electromagnetic waves travel at _____ speed in _____.
All electromagnetic waves are _____ waves.

01.3 Explain why our eyes cannot see all electromagnetic waves.　[2 marks]

01.4 Give **one** use of infrared waves.　[1 mark]

01.5 Describe the dangers of using X-rays and a precaution that a radiographer might take in a hospital.　[2 marks]

> Ultraviolet radiation increases your risk of skin cancer.
> Most sunscreens with an SPF of 15 or higher are good at protecting your skin from the UV rays of the sun. If it takes 20 minutes for your unprotected skin to burn, then using a SPF 15 sunscreen theoretically prevents sunburn for 5 hours. There are problems with the SPF model. No sunscreen, regardless of strength, should be expected to stay effective for longer than two hours without reapplication.

01.6 Evaluate whether using an SPF 50 sunscreen twice a day is suitable for someone with a very fair skin.　[3 marks]

01.7 Describe how radio waves can be produced by oscillations in electrical circuits.　[2 marks]

02.1 Copy and complete the diagram in Figure 1, showing the magnetic field lines on the two magnets.　[3 marks]

Figure 1

02.2 A student finds an old bar magnet but the markings have rubbed off. Describe how the student can find the north seeking pole of the magnet without using another magnet.　[3 marks]
A long piece of wire is wrapped around an iron core. The ends of the wire are attached to a switch and power pack.

Figure 2

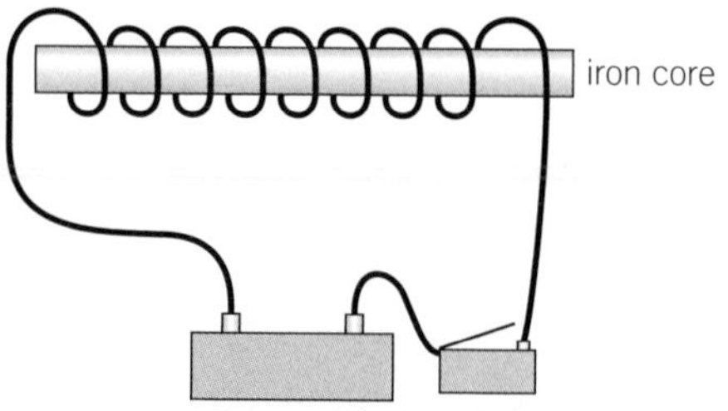

iron core

power pack　　switch

02.3 Give the name of this equipment　[1 mark]

02.4 Give the advantage of wrapping the wire around a soft iron core compared to a wooden rod.　[1 mark]

02.5 Sketch the pattern of the magnetic field lines in and around a coil of wire when there is a current flowing through the coils of wire.　[2 marks]

02.6 The wire used is covered in insulation material. Explain why.　[2 marks]

03 In March 2014 a spacecraft named Rosetta set off to reach a comet deep in our solar system. In November 2014, Rosetta dropped Philae, the first spacecraft to land on a comet, onto the comet. The gravity on the comet is approximately 1/200000 of the gravity on Earth.

03.1 Suggest **one** reason why the spacecraft took so long to reach the comet.　[1 mark]

03.2 Calculate the weight of Rosetta and Philae just before leaving Earth.
Their combined mass was 2900 kg. Gravitational field strength on Earth is 9.8 N/kg.　[2 marks]

03.3 Explain why Philae, 22.5 km above the comet, took seven hours to reach the surface of the comet after leaving the Rosetta spacecraft.　[2 marks]

03.4 Calculate the average speed of descent of Philae. Give your answer in m/s.　[3 marks]

03.5 Some people think that space travel is a waste of money. Do you agree? Give a reason for your answer.　[1 mark]

04 A manufacturer of golf clubs investigated the distance a golf ball would travel using a new type of golf club. They used an automatic golf swing machine and carried out the investigation on the same day at the same location.

Figure 3

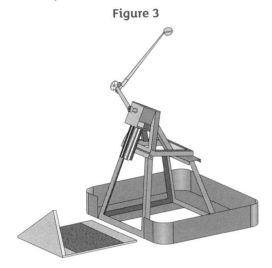

04.1 Why was it important to keep certain factors in the experiment the same? [1 mark]

04.2 Suggest a reason why the investigation was checked by an independent scientist. [1 mark]

04.3 🅗 Calculate the total momentum of a golf ball and club just before impact. The mass of the club head is 0.195 kg and velocity of the club at impact is 54 m/s. [2 marks]

04.4 A laser range finder contains a transmitter and receiver of laser light. Describe how a laser range finder could be used to determine the distance to an object on a golf course. [3 marks]

04.5 Suggest a safety precaution that should be undertaken when laser light is used. [2 marks]

05.1 The normal speed limit in built up areas is 30 mph (13.4 m/s). Give **two** reasons why this speed limit is reduced to 20 mph (8.9 m/s) near schools. [2 marks]

05.2 A road safety council official wants drivers who exceed the 20 mph speed limit near schools to be penalised more harshly than other speeding drivers. Do you agree? Give reasons for your answer. [2 marks]

05.3 Give **two** road safety features, other than speed reduction, that can be found near schools. [2 marks]

05.4 A 1000 kg car hits a stationary object when it is accelerating at 2 m/s². Calculate the impact force. [2 marks]

06 A student investigated how changing the grade of sand paper affected the friction force of the sand paper.

Figure 4

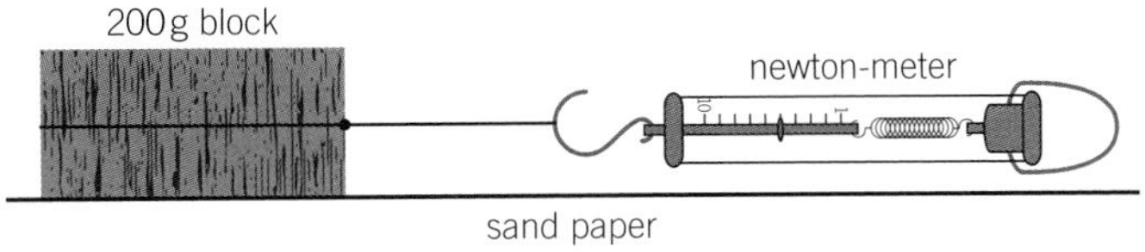

200 g block — newton-meter — sand paper

The student pulled on the newton-meter and measured the force needed to start moving the block of wood. The student recorded the results in a table.

Grade of sand paper	1	2	3	4	5
Force in N	7.5	3.5	2.0	1.5	0.5

06.1 Name a variable that the student controlled. [1 mark]

06.2 What is the resolution of the newton-meter? [1 mark]

06.3 Identify any trends or patterns in the results of the investigation. [2 marks]

06.4 Describe how the results the student obtained can be made reproducible. [2 marks]

06.5 The student wanted to draw a line graph of her results. Give **one** reason why she had to draw a bar chart. [1 mark]

07 The table gives some information about the International Space Station (ISS), which is a manned spacecraft in orbit around the Earth.

Orbital height above the Earth	400 km
Speed of space station	7.6 km/s
Time of one complete orbit	92.6 mins

07.1 Calculate how many complete orbits the ISS will make in one day. [2 marks]

07.2 Calculate the total distance the ISS will travel in one day. [2 marks]

07.3 Explain why the ISS stays at constant speed as it orbits the Earth. You should refer to Newton's 1st law of motion. [2 marks]

08.1 A large painting hangs in a gallery. Draw a scale diagram and determine the weight of the painting. [3 marks]

Figure 5

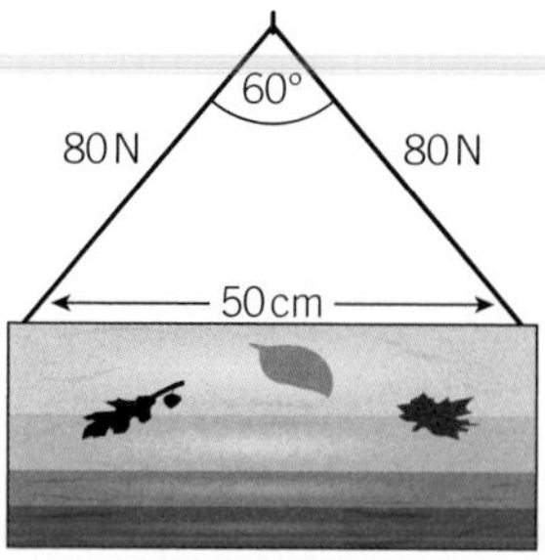

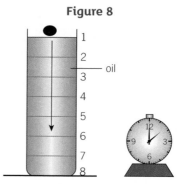

08.2 Figure 6 shows a large box on a slope.

Figure 6

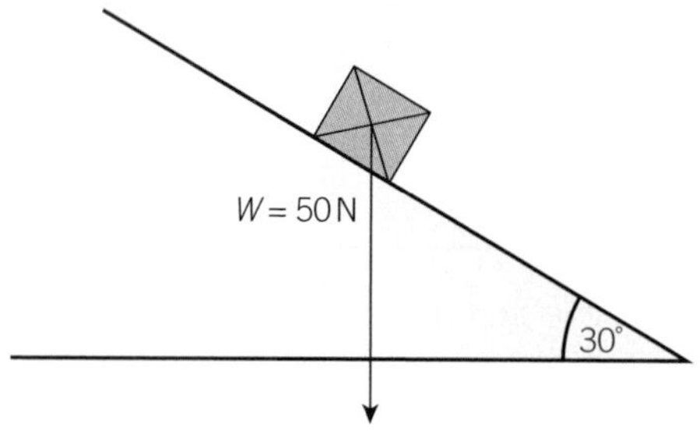

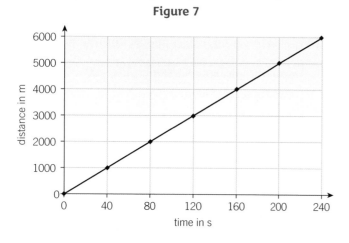

Calculate the mass of the box. Gravitational field strength is 9.8 N/kg [2 marks]

08.3 Draw a scale diagram to resolve the horizontal and vertical components of the weight of the box on the slope. [3 marks]

09.0 The graph of a car journey is shown in Figure 7.

Figure 7

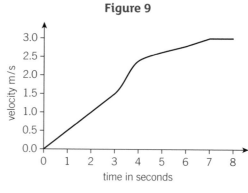

09.1 Calculate the average speed of the car during the journey. [3 marks]

09.2 Calculate the work done by the car during the first 80 s. The driving force of the car is 12 000 N. Give your answer in kJ [3 marks]

09.3 A second car starts from rest and travels 3000 m. The velocity of the car at the end is 50 m/s. Calculate the average acceleration of the car. Give the units. [4 marks]

final velocity^2 − initial velocity^2 = 2 × acceleration × distance

10.0 A student investigates how changing the size of a ball affects the rate of descent of the ball in oil.

Figure 8

10.1 Name the dependent variable. [1 mark]

10.2 Name **two** factors that must remain the same. [2 marks]

10.3 A mechanical clock is used to time the ball. Suggest one improvement. [1 mark]
Figure 9 shows a graph of one set of results.

Figure 9

10.4 Use the graph to calculate the acceleration at 4.5 s. [3 marks]

10.5 Calculate the distance travelled by the ball in the first 2 seconds. [2 marks]

10.6 Describe the shape of the graph in terms of the forces acting on the ball. [4 marks]

11.0 Figure 10 shows the equipment used to investigate the force on a conductor in a magnetic field.

Figure 10

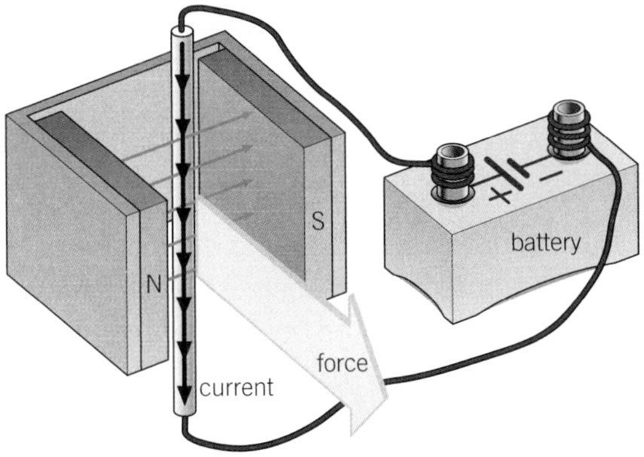

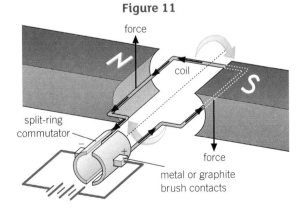

11.1 Which of the following would increase the force?
[2 marks]

A Use a stronger magnet.
B Make the conductor shorter.
C Use a bigger magnet.
D Use an alternating current.
E Increase the potential difference.

11.2 Explain using Fleming's left-hand rule why the direction of the force arrow is correct on the diagram. [3 marks]

11.3 Calculate the magnetic flux density from the information given. Give the units. [4 marks]
● The force on the conductor is 0.0052 N.
● The length of the conductor in the field is 30 mm.
● The current through the conductor is 250 mA.
force = magnetic flux density × current × length

12.0 Figure 11 shows a diagram of a simple electric motor.

Figure 11

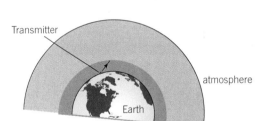

12.1 Choose words from the box to complete the sentences below.

speed	reduced	direction	magnet
reversed	increased	faster	

The coil will turn _____ when the current is _____.
The coil will reverse _____ when the _____ or current is _____. [5 marks]

12.2 Explain how the force on the conductor in a magnetic field causes the rotation of the coil in an electric motor. Your answer should include the purpose of the split rings. [2 marks]

13.0 A transmitter on Earth sends a signal to a satellite in Space. The speed of electromagnetic waves in the atmosphere increases slightly with increased height.
The satellite is in space where there is no atmosphere.

Figure 12

satellite

Transmitter

atmosphere

Earth

13.1 Name the electromagnetic wave used to communicate with satellites. [1 mark]

13.2 Copy and complete the diagram to show the signal travelling from the transmitter to the satellite. [2 marks]

13.3 Give the name of the process that occurs as the signal travels through the atmosphere. [1 mark]

13.4 Complete this sentence.
In addition to its change of speed, the wave signal changes _____ as it travels from one medium to another medium. [2 marks]

13.5 Calculate the speed of microwaves in space. The frequency of the wave is 500 MHz and the wavelength is 60 cm. [3 marks]

Maths skills for Physics
MS1 Arithmetic and numerical computation

Learning objectives

After this topic, you should know how to:

- recognise and use expressions in decimal form
- recognise and use expressions in standard form
- use ratios, fractions and percentages
- make estimates of the results of simple calculations.

Figure 1 *How far away is the Moon?*

Figure 2 *The air pressure at the summit of Mount Everest is significantly lower than at sea level*

What is the speed of a radio wave? How far away is the Moon? What is the difference in air pressure between sea level and the summit of Mount Everest?

Scientists use maths all the time – when collecting data, looking for patterns, and making conclusions. This chapter includes all the maths for your GCSE Physics course. The rest of the book gives you many opportunities to practise using maths in physics.

1a Decimal form

There will always be a whole number of protons, neutrons, or electrons in an atom.

When you make measurements in science the numbers may *not* be whole numbers, but numbers *in between* whole numbers. These are numbers in decimal form, for example 3.2 cm, or 4.5 g.

The value of each digit in a number is called its place value.

Thousands	Hundreds	Tens	Units	·	Tenths	Hundredths	Thousandths
4	5	1	2	·	3	4	5

1b Standard form

Place value can help you to understand the size of a number, however some numbers in science are too large or too small to understand when they are written as ordinary numbers. For example, distance from the Earth to the Sun, 150 000 000 000 m or the diameter of the nucleus of a hydrogen atom, 0.000 000 000 000 001 75 m.

We use standard form to show very large or very small numbers more easily.

In standard form, a number is written as $A \times 10^n$.

- A is a decimal number between 1 and 10 (but not including 10), for example 1.5.
- n is a whole number. The power of ten can be positive or negative, for example 10^{11}.

This gives you a number in standard form, for example, 1.5×10^{11} m.

Table 1 explains how you convert numbers to standard form.

Table 1 *How to convert numbers into standard form*

The number	The number in standard form	What you did to get to the decimal number	...so the power of ten is...	What the *sign* of the power of ten tells you
1000 m	1.0×10^3 m	You moved the decimal point 3 places to the *left* to get the decimal number	+3	The positive power shows the number is *greater* than one.
0.01 s	1.0×10^{-2} s	You moved the decimal point 2 places to the *right* to get the decimal number	−2	The negative power shows the number is *less* than one.

It is much easier to write some of the very big or very small numbers that you find in real life using standard form. For example

- the distance from the Earth to the Sun is 150 000 000 000 m = 1.5×10^{11} m
- the diameter of an atom is 0.000 000 000 1 m = 1.0×10^{-10} m
- the wavelength of light of is around 0.000 0007 m = 7×10^{-7} m
- the speed of light is 300 000 000 m/s = 3×10^8 m/s.

Multiplying numbers in standard form

You can use a scientific calculator to calculate with numbers written in standard form. You should work out which button you need to use on your own calculator (it could be **EE**, **EXP**, **10ˣ**, or **×10ˣ**).

Worked example: Standard form

A train travelled a distance of 180 km at a constant speed in a time of 235 minutes

1 Calculate the time taken in seconds and write your answer in standard form.

2 Calculate the distance travelled by the train each second. Write your answer in standard form to 3 significant figures.

Solution

1 **Step 1:** Because there are 60 seconds in 1 minute, multiply the time in minutes by 60 to give the time in seconds.

Time in seconds = $235 \times 60 = 14\,100$ s.

Step 2: Convert the numbers to standard form

$14\,100$ s = 1.41×10^4 s.

2 **Step 1:** Distance travelled each second = $\dfrac{180\,\text{km}}{1.41 \times 10^4\,\text{s}}$

= 0.012 766 km

Step 2: Convert the numbers to standard form

0.012 766 km = 1.2766×10^{-2} km

Step 3: Write the numbers to 3 significant figures

1.2766×10^{-2} km = 1.28×10^{-2} km

Figure 3 *The Sun is 1.5×10^{11} m from the Earth*

Figure 4 *Uranium isotopes are used in nuclear power plants. The atomic radius of a uranium atom is 1.75×10^{-10} m*

Study tip

Check that you understand the power of ten, and the sign of the power.

Figure 5 *You can use a scientific calculator to do calculations involving standard form*

Study tip

In step 3, the third significant figure is rounded up to 8 because the fourth significant figure is greater than or equal to 5.

1c Ratios, fractions, and percentages
Ratios

A ratio compares two quantities. A ratio of 2:4 of radioactive atoms to non-radioactive atoms in a radioactive sample means for every two radioactive atoms there are four non-radioactive atoms.

You can compare ratios by changing them to the form $1:n$ or $n:1$.

1 : n   Divide both numbers by the *first* number.
For every one radioactive atom there are two non-radioactive atoms.

$$\div 2 \left(\begin{array}{c} 2:4 \\ 1:2 \end{array}\right) \div 2$$

n : 1   Divide both numbers by the *second* number.
For every half a radioactive atom there is one non-radioactive atom (even though you can't really get 'half a radioactive atom').

$$\div 4 \left(\begin{array}{c} 2:4 \\ 0.5:1 \end{array}\right) \div 4$$

You can describe the number of radioactive atoms in relation to the number of non-radioactive atoms using three different ratios 2:4, 1:2 and 0.5:1. All of the ratios are equivalent – they mean the same thing.

You can simplify a ratio so that both numbers are the lowest whole numbers possible.

Worked example: Simplifying ratios

A student draws a vector arrow of length 120 mm to represent a force X of 60 N on a scale diagram. She then needs to draw a second vector arrow to represent a force Y of 36 N on the same diagram.

1 Calculate the ratio of force Y to force X.

2 Use your answer to **1** to calculate the required length of the vector arrow for Y.

Solution

1 Divide force Y by force X

$$\frac{\text{force Y}}{\text{force X}} = \frac{36\,\text{N}}{60\,\text{N}} = 0.60$$

2 Because the diagram is a scale diagram, the ratio of the vector arrow lengths is the same as the ratio of the force.

The length of the vector arrow for Y = 0.60 × the length of the vector arrow for X

$$= 0.60 \times 120\,\text{mm}$$

$$= 72\,\text{mm}$$

Fractions

A fraction is a part of a whole.

$\frac{1}{3}$ → The numerator tells you how many parts of the whole you have.

→ The denominator tells you how many equal parts the whole has been divided into.

To convert a fraction into a decimal, divide the numerator by the denominator.

$\frac{1}{3} = 1 \div 3 = 0.33333\ldots = 0.\dot{3}$ (the dot shows that the number 3 recurs, or repeats over and over again).

To convert a decimal to a fraction, use the place value of the digits, then simplify. $0.045 = \frac{45}{1000} = \frac{9}{200}$

Take care if you are asked to use ratios to find a fraction. For example in a fitness run in which the runner walks then jogs alternately, a ratio of 2:4 in the walking to jogging distance does not mean that the walking distance is $\frac{2}{4}$ (or half) of the total distance. If the total distance was 6 km, the walking distance would be 2 km and the jogging distance would be 4 km. So the walking distance would be $\frac{1}{3}\left(=\frac{2\,\text{km}}{6\,\text{km}}\right)$ of the total distance.

Figure 6 *One square of this chocolate bar represents $\frac{1}{24}$. A column of four squares represents $\frac{4}{24}$, which can be simplified to $\frac{1}{6}$*

Worked example: Calculating the fraction of a quantity

A satellite in a circular orbit above the Earth's equator at a certain height takes 24 hours to orbit the Earth once. During that time, it takes six hours to go over the Atlantic Ocean.

Figure 7 shows a satellite in a circular orbit.

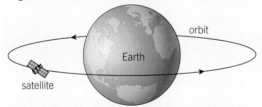

Figure 7 *A satellite in a circular orbit*

1 Calculate the fraction of one complete orbit which is over the Atlantic Ocean and write this down as a decimal.

2 The distance travelled by the satellite in one orbit = 4.0×10^4 km. Calculate the distance travelled by the satellite when it is over the Atlantic Ocean.

Solution

1 Divide the time taken to go over the Atlantic Ocean by the time taken for one complete orbit:

$$\frac{\text{time taken over the Atlantic Ocean}}{\text{time taken for one compete orbit}} = \frac{6}{24} = 0.25$$

2 Distance travelled by the satellite above the ocean

$$= 0.25 \times 4.0 \times 10^4\,\text{km} = 1.0 \times 10^4\,\text{km}$$

Percentages

A percentage is a number expressed as a fraction of 100. For example:

$$77\% = \frac{77}{100} = 0.77$$

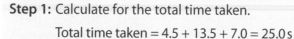

Worked example: Calculating a percentage

A car accelerates from rest for 4.5 s then travels at a constant speed for 13.5 s before braking to a standstill in 7.0 s. Calculate the percentage of time the car travels at constant speed.

Solution

Step 1: Calculate for the total time taken.

Total time taken = 4.5 + 13.5 + 7.0 = 25.0 s

Step 2: Divide the time taken at constant speed by the total time taken and express the answer as a decimal.

$$\frac{\text{time taken at constant speed}}{\text{total time taken}} = \frac{13.5}{25} = 0.54$$

Step 3: Multiply the answer above by 100

$$0.54 \times 100 = \mathbf{54\%}$$

Figure 8 *What percentage of the time the car travelled was at a constant speed?*

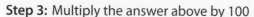

Worked example: Calculating a percentage change

A spring has an unstretched length of 300 mm. Calculate the percentage change in its length when it is stretched to a length of 396 mm.

Solution

Step 1: Calculate the actual change in length.

396 mm – 300 mm = 96 mm

Step 2: Divide the actual change in length by the unstretched length.

$$\frac{96\,\text{mm}}{300\,\text{mm}} = 0.32$$

Step 3: Multiply the answer to Step 2 by 100%.

$$0.32 \times 100\% = \mathbf{32\%}$$

You may need to calculate a percentage of a quantity. To do this, convert the percentage to a decimal and multiply the quantity by the decimal.

Worked example: Using percentage change

The spring is damaged if it is stretched by more than 35% of its unstretched length. Calculate the maximum length of the spring if it is not to be damaged.

Solution

Step 1: Convert 35% to decimal

$35\% = 0.35$

Step 2: Multiply the unstretched length (300 nm) by 0.35 to give the extension of the spring for no damage.

extension $= 0.35 \times 300\,mm = 105\,mm$

maximum length = original length + extension
$= 300\,mm + 105\,mm = \mathbf{405\,mm}$

Synoptic link

To see more examples of the use of percentages, see Topic P5.5 and Maths skills MS1c.

1d Estimating the result of a calculation

When you use your calculator to work out the answer to a calculation you can sometimes press the wrong button and get the wrong answer. The best way to make sure that your answer is correct is to estimate it in your head first.

Worked example: Estimating an answer

You want to calculate distance travelled and you need to find $34\,m/s \times 8\,s$. Estimate the answer and then calculate it.

Solution

Step 1: Round each number up or down to get a whole number multiple of 10.

34 m/s is about 30 m/s

8 s is about 10 s

Step 2: Multiply the numbers in your head.

$30\,m/s \times 10\,s = 300\,m$

Step 3: Do the calculation and check it is close to your estimate.

Distance $= 34\,m/s \times 8\,s = 272\,m$

This is quite close to 300 so it is probably correct.

Notice that you could do other things with the numbers:

$34 + 8 = 42$ $\qquad$ $\dfrac{34}{8} = 4.3$ $\qquad$ $34 - 8 = 26$

Not one of these numbers is close to 300. If you got any of these numbers, you would know that you needed to repeat the calculation.

Sometimes the calculations involve more complicated equations, or standard form.

When carrying out multiplications or divisions using standard form, you should add or subtract the powers of ten to work out roughly what you expect the answer to be. This will help you to avoid mistakes.

1 A sheet of card has a length of 297 mm, a width of 210 mm and a mass of 25.2 grams. A student cut a strip of even width and length 297 mm from the card. The mass of the strip was 5.6 grams.
   a Calculate the ratio of the mass of the strip to the mass of the original card. [1 mark]
   b The ratio of the width of the strip to the width of the original card is equal to mass ratio calculated in part a. Use mass ratio and the width of the original card to calculate the width of the strip of card. Give your answer to the nearest millimetre. [2 marks]

2 A car and a coach join a motorway at the same junction and travel in opposite directions at constant speed. The car travels at a speed of 30 m/s and the coach travels at a speed of 24 m/s.
   a Calculate the ratio of the speed of the car to the speed of the coach. [2 marks]
   b Calculate how far the car has travelled from the motorway junction when the coach has travelled a distance of 12 km from the junction. [2 marks]

MS2 Handling data

2a Significant figures

Numbers are rounded when it is not appropriate to give an answer that is too precise.

When rounding to significant figures, count from the first non-zero digit.

These masses each have three significant figures (s.f.). The significant figures are underlined in each case.

153 g   0.153 g   0.00153 g

Table 1 below shows some more examples of measurements given to different numbers of s.f.

Table 1 *The number of significant figures – the significant figures in each case are underlined*

Number	0.05 s	5.1 nm	0.775 g/s	23.50 cm³
Number of significant figures	1	2	3	4

In general, you should give your answer to the same number of s.f. as the as the data in the question that has the lowest number of s.f.

Remember that rounding to s.f. are **not** the same as decimal places. When rounding to decimal places, count the number of digits that follow the decimal point.

Worked example: Significant figures

Calculate the average speed of an eagle that dives a distance of 230 m in 3.4 s.

Solution

Step 1: Write down what you know.

distance = 230 m (this number has 2 s.f.)

time = 3.4 s (this number also has 2 s.f.)

Step 2: Write down the equation that links the quantities you know and the quantity you want to find.

$$\text{average speed (m/s)} = \frac{\text{distance (m)}}{\text{time (s)}}$$

Step 3: Substitute values into the equation and do the calculation.

$$\text{speed} = \frac{230 \, \text{m}}{3.4 \, \text{s}}$$

$$= 67.647\,058\,823\,529 \, \text{m/s (too many s.f.)}$$

$$= \mathbf{68 \, m/s} \text{ to 2 s.f. (the question uses 2 significant figures, so give your answer to 2 s.f.)}$$

Learning objectives

After this topic, you should know how to:

- use an appropriate number of significant figure
- find arithmetic means
- construct and interpret frequency tables and bar charts
- make order of magnitude calculations.

Figure 1 *What is the average speed of the eagle?*

2b Finding arithmetic means

To calculate the mean value of a series of values:

- add together all the values in the series to get a total
- divide the total by the number of values in the data series

---

Worked example: Calculating a mean

A student's reaction time was measured and recorded five times. Her results were as follows:

$$0.46\,s \qquad 0.48\,s \qquad 0.52\,s \qquad 0.44\,s \qquad 0.56\,s$$

Calculate the mean value of these measurements.

Solution

Step 1: Add together the recorded values.

$$0.46\,s + 0.48\,s + 0.52\,s + 0.44\,s + 0.56\,s = 2.46\,s$$

Step 2: Divide by the number of recorded values (in this case, five measurements were measured).

$$\frac{2.46}{5} = 0.49\,s \text{ (2 s.f.)}$$

The mean reaction time for the student was 0.49 s (2 s.f.)

2c Frequency tables and bar charts
Tables

Tables are used to present data from observations and measurements that are made during experiments or from surveys. Data can be

- qualitative (descriptive, but with no numerical measurements)
- quantitative (including numerical measurements).

Qualitative data includes categoric variables, such as types of power stations. The values of categoric variables are names not numbers.

Quantitative data includes

- continuous variables – can take any value (usually collected by measuring), such as length or time.
- discrete variables – can only take exact values (usually collected by counting), such as number of paper clips picked up by an electromagnet, or complete swings of a pendulum.

Table 2 below shows the energy output in 2010 of different types of UK power stations as a percentage of the total UK energy output.

- The first column is a categoric variable as it shows the different types of power stations.
- The second column is a continuous variable as it shows the energy output of each type of power station as a percentage of the total energy output.

Table 2 *The percentage energy output of different types of electricity power stations in the UK in 2010*

Type of power station	% energy output
Gas	40
Coal	31
Oil	1
Nuclear	18
Wind	3
Hydroelectric	2
Bio-energy	3
Other	2

Sometimes data is grouped into classes, as in Table 3, which shows the ages of the winners of the Nobel Prize for Physics. If you need to group data into classes:

● make sure the values in each class do not overlap

● aim for a sensible number of classes.

Table 3 *Ages of the winners of the Nobel Prize for Physics*

Ages, years	Number of physicists
21–30	1
31–40	28
41–50	55
51–60	51
61–70	35
71–80	23
81–90	8

Figure 2 *William Lawrence Bragg is the youngest winner of the Nobel Prize for Physics. He was 25 when he was jointly award the prize with his father, William Henry Bragg, in 1915*

Bar charts

You can use a bar chart to display data.

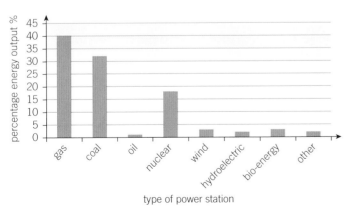

Figure 3 *The percentage energy output of different types of electricity power stations in the UK in 2010*

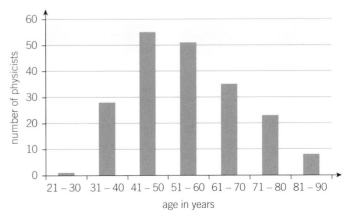

Figure 4 *Ages of the winners of the Nobel Prize for Physics*

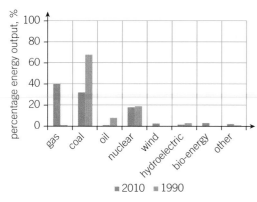

Figure 5 *The different types of electricity power stations in the UK and their percentage power output in 2010 and 1990*

Figure 6 *What is the median speed of a group of horses?*

You can also use bar charts to compare two or more sets of data (Figure 5)

2d Averages

When you collect data, it is sometimes useful to calculate an average. There are three ways you can calculate an average – the mean, the mode, and the median.

You saw how to calculate a mean earlier in this topic (section 2b).

How to calculate a median

When you put the values of a series in order from smallest to biggest the middle value is called the median. When the series of values has two central values, the median is the mean of these two values.

Worked example: Calculating a median (odd number of values)

The speed of seven different horses galloping is shown below.

11.1 m/s   13.2 m/s   11.9 m/s   12.5 m/s   12.8 m/s   11.2 ms   12.3 m/s

Calculate the median speed of the horses.

Solution

Step 1: Place the values in order from smallest to largest:

11.1 m/s   11.2 m/s   11.9 m/s   12.3 m/s   12.5 m/s   12.8 m/s   13.2 m/s

Step 2: Select the middle value – this is the median value.

median value = 12.3 m/s

If you have an even number of values, you select the middle pair of values and calculate a mean. That then becomes your median value.

How to calculate a mode

The mode is the value that occurs most often in a series of results. If there are two values that are equally common, then the data is bimodal.

Worked example: Calculating a mode

A student measured the time it took for 12 different cars to travel a 100 m stretch of road.

12.9 m/s   13.1 m/s   17.3 m/s   14.5 m/s   13.0 m/s   13.1 m/s

12.9 m/s   13.2 m/s   13.4 m/s   12.5 m/s   14.1 m/s   12.9 m/s

Calculate the modal time for the cars to travel the stretch of road.

Solution

Step 1: Place the values in order from smallest to largest.

12.5 m/s 12.9 m/s 12.9 m/s 12.9 m/s 13.0 m/s 13.1 m/s
13.1 m/s 13.2 m/s 13.4 m/s 14.1 m/s 14.5 m/s 17.3 m/s

Step 2: Select the value which occurs the most often.

mode = 12.9 m/s

2e Scatter diagrams and correlations

You may collect data and plot a scatter graph (see Topic M4). You can add a line to show the trend of the data, called a line of best fit. The line of best fit is a line that goes through as many points as possible and has the same number of points above and below it.

If the gradient of the line of best fit is:

positive it means as the independent variable *increases* the dependent variable *increases*

negative it means as the independent variable *increases* the dependent variable *decreases*

zero it means changing the independent variable has no effect on the dependent variable.

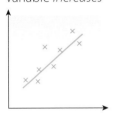 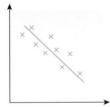 

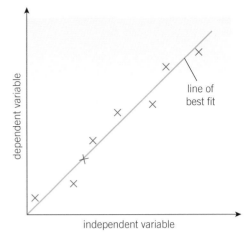

Figure 7 *A scatter graph*

A relationship where there happens to be a link is called a correlational relationship, or correlation. You say that the relationship between the variables is positive, negative, or that there is no relationship.

For example:

● As you increase the force on an object, the acceleration of the object will increases – a positive correlation.

● As you increase the density of a liquid, the speed with which an object falls through the liquid will decrease – a negative correlation.

● The intensity of a light beam has no effect of the angle of refraction in a glass block – no relationship.

The presence of a relationship does not always mean that changing the independent variable *causes* the change in the dependent variable. In order to claim a causal relationship, you must use science to predict or explain *why* changing one variable affects the other.

Often there is a third factor that is common to both so it looks as if they are related. You could collect data for shark attacks and ice cream sales. A graph shows a positive correlation but shark attacks do not make people buy ice cream. Both are more likely to happen in the summer.

2f Order of magnitude calculations

Being able to make a rough estimate is helpful. It can help you to check that a calculation is correct by knowing roughly what you expect the answer to be. A simple estimate is an order of magnitude estimate, which is an estimate to the nearest power of 10.

For example, to the nearest power of 10, you are probably 1 m tall and can walk at a speed of about 1 m/s.

You, your desk, and your chair are all of the order of 1 m tall. The diameter of a molecule is of the order of 1×10^{-9} m, or 1 nanometre.

1 A student used an electronic stopwatch to measure how long a ball took to roll a measured distance down a slope after being released. She repeated the measurement 10 times. Her results were as follows:

1.70 s	1.65 s	1.82 s	1.61 s	1.74 s
1.61 s	1.63 s	1.71 s	1.67 s	1.65 s

  a Calculate the mean value of these measurements. Give your answer to 3 significant figures. [1 mark]

  b Calculate the median value of these measurements. [1 mark]

  c Calculate the mode value of these measurements. [1 mark]

2 How many significant figures are the following numbers quoted to?

  a 0.001 [1 mark]   b 85.4 [1 mark]

  c 70.0 [1 mark]   d 8.314 [1 mark]

  e 3.08×10^8 [1 mark]   f 3.07×10^{-3} [1 mark]

3 Look back at Figure 5. The bar chart tells you about the types of power stations used to generate electricity in the United Kingdom.

  a For each non-renewable power station, write down one conclusion you draw about their contribution to UK electricity generation in 2010 compared with 1990. [4 marks]

  b Give two conclusions you draw from the chart about the use of renewable energy sources compared with the use of fossil fuel in

    i 1990 [2 marks]   ii 2010? [2 marks]

4 Write down an order of magnitude value for the speed of an ant and estimate the distance it could travel in 1 minute. [2 marks]

MS3 Algebra

3a Mathematical symbols

You have used lots of different symbols in maths, such as $+$, $-$, $\times$, $\div$. There are other symbols that you might meet in science. These are shown in Table 1.

Table 1 *The symbols you will meet whilst studying physics*

Symbol	Meaning	Example
$=$	Equal to	$2\,m/s \times 2\,s = 4\,m$
$<$	Is less than	The mean weight of a child in a family $<$ the mean weight of an adult in a family
$<<$	Is very much less than	The diameter of an atom $<<$ the diameter of an apple
$>>$	Is very much bigger than	The diameter of the Earth $>>$ the diameter of a football
$>$	Is greater than	The density of lead $>$ the density of steel
$\propto$	Is proportional to	F(force) $\propto$ x (extension) for a spring
$\sim$	Is approximately equal to	$272\,m \sim 300\,m$

3b Changing the subject of an equation

An equation shows the relationship between two or more variables. You can change an equation to make *any* of the variables become the subject of the equation.

To change the subject of an equation, you can do an opposite (inverse) operation to both sides of the equation to get the variable that you want on its own.

This means that:

- subtracting is the opposite of adding (and adding is the opposite of subtracting)
- dividing is the opposite of multiplying (and multiplying is the opposite of dividing)
- taking the square root is the opposite of squaring (and squaring is the opposite of taking the square root)

You can use these steps to change the subject of an equation, such as the equation for kinetic energy.

Learning objectives

After this topic, you should know how to:

- understand and use the symbols:
  $=$, $<$, $<<$, $>>$, $>$, $\propto$, $\sim$
- change the subject of an equation
- substitute numerical values into algebraic equations using appropriate units for quantities.

Synoptic link

To see more examples of rearranging equations, see Topic P2.2 and Topic P6.5.

Figure 1 *If you know the kinetic energy and mass of this roller coaster car you can work out the speed*

Worked example: Kinetic energy

Change the equation $KE = \frac{1}{2}mv^2$ to make v the subject.

Solution

Step 1: Multiply both sides of the equation by 2.

$$2 \times KE = 2 \times \frac{1}{2}mv^2$$

Remove the fraction from the right hand side of the equation: $2 \times KE = mv^2$

Step 2: Divide by m to get the v^2 on its own.

$$\frac{2 \times KE}{m} = \frac{mv^2}{m} = \frac{2 \times KE}{m} = v^2$$

Step 3: Take the square root of both sides.

$$\sqrt{v^2} = \sqrt{\frac{2 \times KE}{m}} = v = \sqrt{\frac{2 \times KE}{m}}$$

3c Quantities and units
SI Units

When you make a measurement in science you need to include a number *and* a unit.

When you do a calculation your answer should also include both a number *and* a unit. There are some special cases where the units cancel, but usually they do not.

Everyone doing science, including you, needs to use the SI system of units.

Table 2 *Some quantities and their units you will meet during your physics GCSE*

Quantity and symbol	Unit
distance *s*	metre m
mass *m*	kilogram kg
time *t*	second s
current *I*	ampere A
temperature *t*	kelvin K
frequency *f*	hertz Hz
force *F*	newton N
energy *E*	joule J
power *P*	watt W
pressure *p*	pascal Pa
charge *Q*	coulomb C
potential difference *V*	volt V
electric resistance *R*	ohm Ω

For example, 1.5 N is a *measurement*. The number 1.5 is not a measurement because it does not have a unit.

Some quantities that you *calculate* do not have a unit because they are a ratio.

Using units in equations

When you put quantities into an equation it is best to write the number *and* the unit. This helps you to work out the unit of the quantity that you are calculating.

Worked example: Speed $= \dfrac{\text{distance}}{\text{time}}$

A sprinter can run 100 m in 10 s. Calculate the average speed of the sprinter.

Solution

Step 1: Write down what you know.

distance   =   100 m

time   =   10 s

Step 2: Write down the equation you need.

$$\text{average speed (m/s)} = \dfrac{\text{distance (m)}}{\text{time (s)}}$$

Step 3: Do the calculation and include the units.

$$\text{average speed} = \dfrac{100 \text{ m}}{10 \text{ s}} = \mathbf{10 \text{ m/s}}$$

m/s are the units of speed.

Figure 2 *If you know the distance a sprinter ran, and the time it took him, you can calculate his average speed*

Metric prefixes

You can use metric prefixes to show large or small multiples of a particular unit. Adding a prefix to a unit means putting a letter in front of the unit, for example km. It shows you that you should multiply your value by a particular power of 10 for it to be shown in an SI unit.

For example, 3 millimetres $= 3 \text{ mm} = 3 \times 10^{-3} \text{ m}$. Most of the prefixes that you will use in physics involve multiples of 10^3. However, when dealing with volumes in density calculations, we often deal with cubic centimetres (cm^3), where 1 cm is one hundredth of a metre (or $1 \times 10^{-2} \text{ m}$).

Table 3 *Common prefixes you will use in your units*

Prefix	giga	mega	kilo	deci	centi	milli	micro	nano
Symbol	G	M	k	d	c	m	μ	n
Multiplying factor	10^{9}	10^{6}	10^{3}	10^{-1}	10^{-2}	10^{-3}	10^{-6}	10^{-9}

Converting between units

It is helpful to use standard form when you are converting between units. To do this, it's best to consider how many of the 'smaller' units are contained within one of the 'bigger' units. For example

- There are 1000 mm in 1 m. So $1\text{ mm} = \dfrac{1}{1000}\text{ m} = 10^{-3}\text{ m}$

- There are 1000 m in 1 km. So $1\text{ km} = 1000\text{ m} = 10^{3}\text{ m}$.

1 a Write down the SI symbol for the SI unit of each of the following quantities

    i length [1 mark]   **ii** volume [1 mark]

    iii mass [1 mark]   **iv** force [1 mark]

  b Write down each of the following amount in standard form without the prefix

    i 72 mm [1 mark]   **ii** 16 cm³ [1 mark]

    iii 385 MJ [1 mark]   **iv** 56 kN [1 mark]

2 How would you read the following expressions as a sentence?

  a circumference of the Earth << circumference of the Sun [1 mark]

  b atmospheric pressure ∝ altitude [1 mark]

  c 4.5 kJ ~ 4.61 kJ [1 mark]

3 The force F (N) needed to extend a spring by extension e (m) is given by the equation $F = ke$ where k is the spring constant of the spring.

  a Rearrange the equation to make k the subject. [1 mark]

  b Show that the unit of k is N/m [1 mark]

4 The energy E stored in a stretched spring is given by the equation

  $E = \dfrac{1}{2}ke^2$ where k is the spring constant of the spring and e is its extension.

  a Rearrange the above equation to make e the subject of the equation. [1 mark]

  b What is the unit of e? [1 mark]

5 An object accelerates uniformly for a distance s with an acceleration a. its final velocity v is given by the equation $v^2 = u^2 + 2as$ where $u =$ its initial velocity is u.

  a Rearrange the equation $v^2 = u^2 + 2as$ to make s the subject of the equation. [1 mark]

  b A car accelerates for a certain time with a constant acceleration of 3.4 m/s² from an initial speed of 4.0 m/s to a speed on 22 m//s. Calculate the distance it travels in this time. [1 mark]

MS4 Graphs

During your GCSE course you will collect data in different experiments or investigations. In investigations, the data will either be:

- from an experiment where you have changed *one* independent variable (or allowed time to change) and measured the effect on a dependent variable
- from an investigation where you have collected data about *two* independent variables to see if they are related.

4a Collecting data by changing a variable

In many investigations you change one variable (the independent variable) and measure the effect on another variable (the dependent variable). In a fair test, the other variables are kept constant.

For example, you can vary the amount of force you apply to an object (independent variable) and measure the effect on the acceleration of the object (dependent variable).

A scatter diagram lets you show the relationship between two numerical values.

- The independent variable is plotted on the horizontal axis – the *x*-axis.
- The dependent variable is plotted on the vertical axis – the *y*-axis.

The line of best fit is a line that goes roughly through the middle of all point on the scatter graph. The line of best fit is drawn so that the points are evenly distributed on either side of the line.

Sometimes, graphs are plotted with the independent variable on the *y*-axis because the gradient represents a physical quantity. For example, in an investigation of the motion of an object, the time taken by the object to fall through different distances can be measured. The distance fallen is the independent variable and the time taken is the dependent variable. Plotting the distance on the *y*-axis and time taken on the *x*-axis gives a graph where the gradient of the line represents the speed of the object.

4b Straight line graphs

When people say 'plot a graph' they usually mean plot the points then draw a line of best fit. This is a smooth line that passes through or near each plotted point. If the line of best fit is a straight line, the gradient of the line is constant and there is a linear relationship between the two variables.

Learning objectives

After this topic, you should know how to:

- translate information between graphical and numeric form
- explain that $y = mx + c$ represents a linear relationship
- plot two variables from experimental or other data
- determine the slope and intercept of a linear graph
- draw and use the slope of a tangent to a curve as a measure of rate of change
- describe the physical significance of area between a curve and the *x*-axis and measure it by counting squares as appropriate.

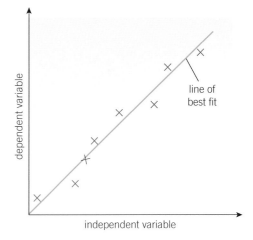

Figure 1 *A graph with the independent variable plotted on the x-axis and the dependent variable plotted on the y-axis*

Study tip

Look back at section MS2e to remind yourself about interpreting scatter graphs and correlations.

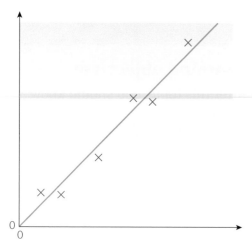

Figure 2 *A line of best fit which passes through the origin*

The equation for a straight line graph

A straight line graph tells you about the mathematical relationship between variables but there are other things that you can calculate from a graph.

The mathematical equation of a straight line is $y = mx + c$, where m is the **gradient** and c is the point on the y-axis where the graph intercepts, called the y-intercept.

Straight line graphs that go through the origin (0,0) are special (Figure 2). For these graphs, y is directly proportional to x, and the mathematical equation is $y = mx$ (because the y-intercept is zero).

When you describe the relationship between two *physical* quantities, you should think about the reason why the graph might (or might not) go through (0,0).

Worked example: Hooke's Law

A spring is being stretched by a force F and experiences an extension x.

A line of best fit for a graph of force (y-axis) against extension (x-axis) is a straight line through (0, 0). Explain what the gradient shows in this context, and explain why the graph goes through (0,0).

Solution

Step 1: Match the equation to $y = mx$ to work out what the gradient means.

$y = mx$, and Hooke's Law says $F = kx$.

So the gradient gives us the value of k, the spring constant.

Step 2: Think about what happens to x when the y quantity is zero.

The line goes through (0,0) because when the force is zero, there is no extension in the spring.

4c Plotting data

When you draw a graph you choose a scale for each axis.

- The scale on the x-axis should be *the same* all the way along the x-axis but it can be *different* to the scale on the y-axis.

- Similarly, the scale on the y-axis should be *the same* all the way along the y-axis but it can be *different* to the scale on the x-axis.

- Each axis should have a label and a unit, such as time / s.

4d Determining the slope and intercept of a straight line

You often need to calculate a gradient. These may represent important physical quantities. Their units can give you a clue as to what they represent. The gradient of a straight line is calculated using the equation:

$$\text{gradient} = \frac{\text{change in } y}{\text{change in } x}$$

For all graphs where the quantity on the x-axis is time, the gradient will tell you the rate of change of the quantity on the y-axis with time.

You can also find the y-intercept of a graph. This is the value of the quantity on the y-axis when the value of the quantity on the x-axis $= 0$.

For all graphs where the quantity on the x-axis is time, the gradient will tell you the rate of change of the quantity on the y-axis with time. For example,

● the rate of change of distance (y-axis) with time (x-axis) is speed,

● the rate of change of velocity (y-axis) with time (x-axis) is acceleration

If the line is straight, its gradient is constant so the rate of change is constant.

4e Using tangents with curved graphs

When you plot a graph of the relationship between certain variables, you may not get a straight line – the relationship is non-linear.

To find the gradient at a point **T** you need to draw a **tangent** to the curve at that point (Figure 3):

● Draw a tangent to the curve – the line should pass through point **T** and have the same slope as the curve at that point.

● Make a right-angled triangle with your line as the hypotenuse. The triangle can be as big as you like , but make sure that the triangle is large enough for you to calculate sensible changes in values.

● Calculate the slope of the tangent using $\dfrac{\text{change in } y}{\text{change in } x}$

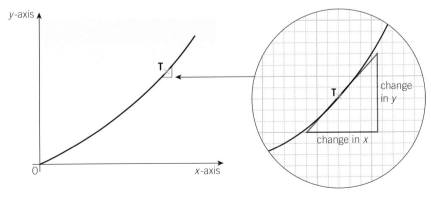

Figure 3 *You find the gradient by drawing a tangent*

If the *x*-axis is time, the gradient will be equal to the rate of change of the variable on the *y*-axis.

On a distance-time graph this is $\frac{\Delta s}{\Delta t}$, or the rate of change of distance with time, or speed.

On a speed-time graph of an object moving in a constant direction, this is $\frac{\Delta v}{\Delta t}$, or the rate of change of speed with time, or acceleration.

4f Measuring the area of graphs

You may also need to find the area of the graph. To find the area you can find the value of one square and then count the squares.

Worked example: Finding the distance travelled
Figure 4 shows a small section of a cyclist's journey.

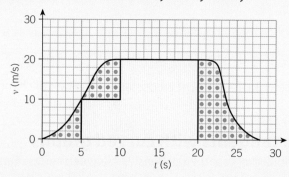

Figure 4 *The cyclist's journey*

Use the graph to calculate an estimate of the distance travelled by the cyclist in this section of her journey.

Solution

Step 1: Calculate the area of any large squares or rectangles contained under the graph (the shaded area).

Small shaded square (from *t* = 5 to *t* = 10), distance represented = 10 m/s × 5 s = 50 m.

Large shaded square (from *t* = 10 to *t* = 20), distance represented = 20 m/s × 10 s = 200 m.

Step 2: Calculate the area of one small square.

The height of each small square is 2 m/s. The width of each small square is 1 s.

So each small square represents 2 m/s × 1 s = 2 m.

Step 3: Count the remaining small squares under the graph that we didn't include in Step 1.

Add half squares together.

Total squares = 64.

Step 4: Multiply by the distance represented by each small square.

Distance = 64 squares × 2 m/square

= 128 m.

Step 5: Add together the distances to find the total distance.

Total distance = 50 m + 200 m + 128 m

= **378 m**

Synoptic link

To see more examples of the use of graphs in physics, see Topics P9.1 and P9.4.

1   **a**   The mathematical equation for a straight line graph is $y = mx + c$. What is represented in this equation by:

  i m [1 mark] **ii** c [1 mark]

  b Sketch a straight line graph with:

  i   a positive gradient and y-intercept (0,0). [2 marks]

  ii a smaller positive gradient than your graph from **i** and a negative y-intercept. [1 mark]

2   The equation $F = k e$ gives the force F applied to stretch a spring so its extension is e. The spring constant of the spring is k. Match this equation to the straight line graph equation $y = mx + c$ [2 marks]

3   For an object that accelerates uniformly from an initial velocity u, its velocity v after time t is given by the equation $v = u + at$. Match the equation to the straight line graph equation $y = mx + c$. [2 marks]

4   Figure 5 shows a distance–time graph.

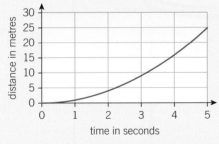

Figure 5 *Distance–time graph for a sprinter*

Calculate the gradient of the line at 4 seconds. [3 marks]

MS5 Geometry and trigonometry

Learning objectives

After this topic, you should know how to:

- use angular measures in degrees
- visualise and represent 2D and 3D forms including two dimensional representations of 3D objects
- calculate areas of triangles and rectangles, surface areas and volumes of cubes.

5a Measuring and using angles

You measure angles with a protractor. Angles are measured in degrees (°). The angle shown in Figure 1 is 45°.

There are 360° in a circle, 180° in a half circle, and 90° in a quarter of a circle.

The three angles of a triangle always add up to 180°. In Figure 1, the angle in the bottom right corner of the triangle is 90°. This is also called a right angle.

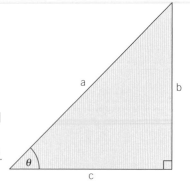

Figure 1 *The symbol for an angle is usually* θ

5b Representation of 3D objects

Three-dimensional (3D) objects have width, height, and depth. Two-dimensional (2D) objects only have width and height. You cannot accurately show all aspects of a 3D object in a diagram. As such, it is easier to draw laboratory apparatus in 2D. Figure 2 and Figure 3 show a beaker drawn in different ways.

600 ml

Figure 2 *A perspective drawing to give a sense of three dimensions*

Figure 3 *A cross-section drawing with no attempt to represent three dimensions*

Other 2D representations of 3D objects include:

- nets – 2D shapes that can be cut out and folded to make 3D shapes (Figure 4)

- elevation views (showing objects from a side)

- plan views (showing objects from above or below).

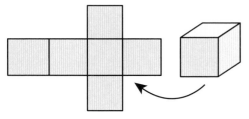

Figure 4 *A cube can be represented as a net*

5c Calculating areas and volumes
Areas of rectangles and triangles

The SI unit of area is the square metre m^2.

Because a distance of 1 metre = 100 cm = 1000 mm, an area of 1 square metre (m^2) = 10 000 cm^2 = 1 000 000 mm^2

A rectangle is a flat shape with four straight sides and a right angle at each of its four corners.

$$\text{area of a rectangle} = \text{height } h \times \text{base } b$$

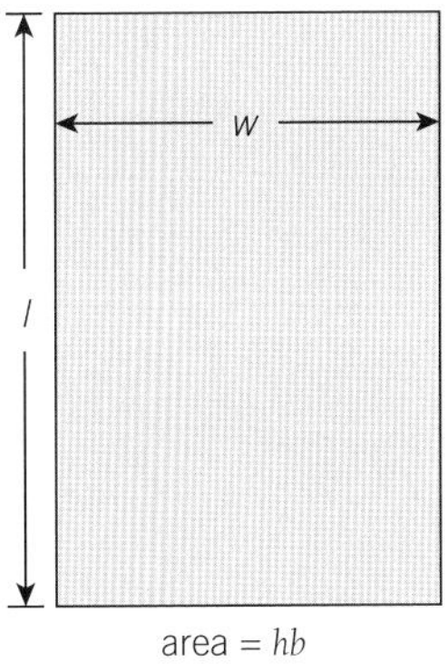

A triangle is a flat shape with three straight sides.

$$\text{area of a triangle} = \frac{1}{2} \times \text{height } h \times \text{base } b$$

area = hb

Figure 5 *Calculating an area of a rectangle*

> **Worked example: Area of a triangle**
>
> Calculate the area of a triangle that has a base of length 0.35 m and a height of 0.12 m. Give your answer in standard form.
>
> **Solution**
>
> area $= \dfrac{1}{2} \times$ its height $\times$ its base $= 0.5 \times 0.12\,\text{m} \times 0.35\,\text{m} = 0.021\,\text{m}^2$
>
> $= 2.1 \times 10^{-2}\,\text{m}^2$

Volumes of cubes

The SI unit of volume is the cubic metre (m^3).

Because a distance of 1 metre = 100 cm = 1000 mm,

a volume of 1 cubic metre (m^3) = 1 000 000 cm^3 = 1 000 000 000 mm^3

A cuboid is a rectangular box with sides of unequal length (Figure 7).

$$\text{volume of a cuboid} = \text{length } l \times \text{width } w \times \text{height } h$$

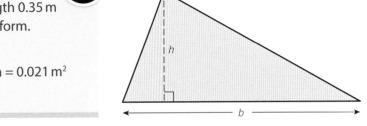

area = $\frac{1}{2}hb$

Figure 6 *Calculating an area of a triangle*

> **Synoptic link**
>
> To see more examples of calculating areas, see Maths skill MS5c.

> **Worked example: Volume of a cuboid**
>
> Calculate the volume of a cuboid that has sides of length 8.0 cm, height 5.0 cm and width 3.0 cm. Give your answer in cubic metres in standard form.
>
> **Solution**
>
> volume of cuboid = length × width × height
>
> $\qquad$ = 8.0 cm × 5.0 cm × 3.0 cm
>
> $\qquad$ = 0.080 m × 0.050 m × 0.030 m
>
> $\qquad$ = 0.00012 m^3
>
> $\qquad$ = $1.2 \times 10^{-4}\,\text{m}^3$

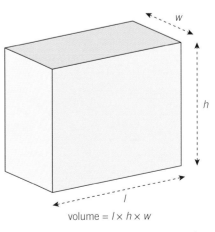

volume = $l \times h \times w$

Figure 7 *Calculating the volume of a cuboid*

A cube is a type of cuboid where the length, height, and width are all the same. The volume of a cube can therefore be simplified to:

$$volume = length^3$$

1 Calculate the area of
   a a rectangle of length 2.3 m and width 1.9 m   [1 mark]
   b a triangle which has a base of 0.35 m and a height
      of 1.2 m.   [1 mark]
   Give your answers to 2 significant figures.

2 Each side of a cube has a length of 1.5 cm.
   a Calculate the volume of the cube in cubic metres.   [1 mark]
   b Calculate the surface area of the cube in square metres.   [1 mark]
   Give your answers to 2 significant figures.

3 The kite in Figure 8 has a length of 1.60 m and a width of 1.15 m.

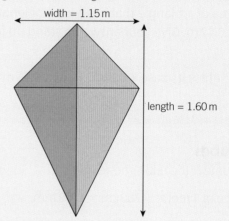

Figure 8

By dividing the area of the kite as a red and a blue triangle, show that the area of the kite is 1.84 square metres.   [3 marks]

Working scientifically

WS1 Development of scientific thinking

Science works for us all day, every day. Working as a scientist you will have knowledge of the world around you, particularly about the subject you are working with. You will observe the world around you. An enquiring mind will then lead you to start asking questions about what you have observed.

Science usually moves forward by slow steady steps. Each small step is important in its own way. It builds on the body of knowledge that we already have. In this book you can find out about:

- how scientific methods and theories change over time (Topics P7.1, P7.2, P7.8, P16.1, P16.4, P16.5)
- the models that help us to understand theories (Chapter 7)
- the limitations of science, and the personal, social, economic, ethical and environmental issues that arise (Topics P2.5, P3.4, P3.5, P7.8, P7.9, P10.7)
- the importance of peer review in publishing scientific results (Topic P7.2)
- evaluating risks in practical work and in technological applications (Topics WS2, P3.4, P3.5, P7.4, P7.6, P7.9).

The rest of this section will help you to 'work scientifically' when planning, carrying out, analysing and evaluating your own investigations.

Figure 1 *All around you, everyday, there are many observations you can make. Studying science can give you the understanding to explain and make predictions about some of what you observe*

WS2 Experimental skills and strategies

Deciding on what to measure

Variables can be one of two different types:

- A categoric variable is one that is best described by a label (usually a word). The type of material used in an experiment is a categoric variable (e.g., copper).
- A continuous variable is one that you measure, so its value could be any number. Temperature (as measured by a thermometer or temperature sensor) is a continuous variable. Continuous variables have values (called quantities). These are found by taking measurements (e.g., mass, volume, etc.) and S.I. units such as grams (g), metres (m), and joules(J) should be used.

Making your data repeatable and reproducible

When you are designing an investigation you must make sure that you, and others, can trust the data you plan to collect. You should ensure that each measurement is *repeatable*. You can do this by getting consistent sets of repeat measurements and taking their mean. You can also have more confidence in your data if similar results are obtained by different investigators using different equipment, making your measurements *reproducible*.

You must also make sure you are measuring the actual thing you want to measure. If you don't, your data can't be used to answer your original question. This seems very obvious, but it is not always easy to set up. You need to make sure that you have controlled as many other variables as you can. Then no-one can say that your investigation, and hence the data you collect and any conclusions drawn from the data, is not valid.

How might an independent variable be linked to a dependent variable?

- The independent variable is the one you choose to vary in your investigation.
- The dependent variable is used to judge the effect of varying the independent variable.

These variables may be linked together. If there is a pattern to be seen (e.g., as one thing gets bigger the other also gets bigger), it may be that:

- changing one has caused the other to change
- the two are related (there is a correlation between them), but one is not necessarily the cause of the other.

Starting an investigation

As scientists, we use observations to ask questions. We can only ask useful questions if we know something about the observed event. We will not have all of the answers, but we will know enough to start asking the correct questions.

When you are designing an investigation you have to observe carefully which variables are likely to have an effect.

An investigation starts with a question and is followed by a prediction, and backed up by scientific reasoning. This forms a hypothesis that can be tested against the results of your investigation. You, as the scientist, predict that there is a relationship between two variables.

You should think about carrying out a preliminary investigation to find the most suitable range and interval for the independent variable.

Making your investigation safe

Remember that when you design your investigation, you must:

- look for any potential hazards
- decide how you will reduce any **risk**.

You will need to write these down in your plan:

- write down your plan
- make a risk assessment
- make a prediction and hypothesis
- draw a blank table ready for the results.

Different types of investigation

A fair test is one in which only the independent variable affects the dependent variable. All other variables are controlled and kept constant.

Figure 2 *Safety precautions should be appropriate for the risk. The wires in electrical circuits may become warm, but you do not need to wear safety gloves. You should, however, let your teacher know if circuit wires begin to heat up*

This is easy to set up in the laboratory, but it can be difficult in outdoor experiments (eg., measuring the speed of sound in air), and is almost impossible in fieldwork. Investigations in the environment are not that simple and easy to control. There are complex variables that are changing constantly.

So how can we set up the fieldwork investigations? The best you can do is to make sure that all of the many variables change in much the same way, except for the one you are investigating. For example, if you are monitoring the effects of aircraft noise on people living near an airport, they should all be experiencing the same noise from other outdoor sources – even if it is constantly changing.

If you are investigating two variables in a large population then you will need to do a survey. Again, it is impossible to control all of the variables. For example, imagine scientists investigating the effect of overhead electricity cables on the health of people living at different distances from the cables. They would have to choose people of the same age and same family history to test. Remember that the larger the sample size tested, the more valid the results will be.

Control groups are used in these investigations to try to make sure you are measuring the variable you intend to measure. For example, when investigating the effect of aircraft noise on people living near an airport, the control groups would use people not living near an airport, but still experiencing the same noise from other outdoor sources as the people living near the airport. The control groups would need to be in similar areas to the airport groups, with similar traffic and other non-airport sources of noise. In this way, the effect on people of living near an airport could be compared with the effect on the control groups.

Designing an investigation
Accuracy
Your investigation must provide accurate data. Accurate data is essential if your results are going to have any meaning.

How do you know if you have accurate data?
It is very difficult to be certain. Accurate results are very close to the true value. However, it is not always possible to know what the true value is.

- Sometimes you can calculate a theoretical value and check it against the experimental evidence. Close agreement between these two values could indicate accurate data.

- You can draw a graph of your results and see how close each result is to the line of best fit.

- Try repeating your measurements and check the spread or range within sets of repeat data. Large differences in a repeated measurement suggest inaccuracy. Or try again with a different measuring instrument and see if you get the same readings.

Precision
Your investigation must provide data with sufficient precision (i.e., close agreement within sets of repeat measurements). If it doesn't then you will not be able to make a valid conclusion.

Figure 3 *Imagine you wanted to investigate the effect of overhead electricity cables on the health of people living at different distances from the cables. You would need to choose a control group using people far away enough from the cables to not be affected by them, but close enough to be still experiencing similar environmental conditions*

Study tip

Trial runs will tell you a lot about how your investigation might work out. They should get you to ask yourself:

- do I have the correct conditions?
- have I chosen a sensible range?
- have I got sufficient readings that are close enough together? The minimum number of points to draw a line graph is generally taken as five.
- will I need to repeat my readings?

Study tip

A word of *caution*.

Just because your results show precision it does not mean your results are accurate.

Imagine you carry out an investigation into the specific heat capacity of a substance. You get readings of the temperature change in the substance that are all the same. This means that your data will have precision, but it doesn't mean that they are necessarily accurate.

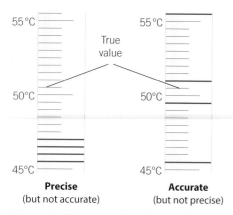

Figure 4 *The green line shows the true value and the pink lines show the readings two different groups of students measured. Precise results are not necessarily accurate results*

Figure 5 *Despite the fact that a stopwatch has a high resolution, it is not always the most appropriate instrument to use for measuring time*

Precision versus accuracy

Imagine measuring the temperature after a set time when a fuel is used to heat a fixed volume of water. Two students repeated this experiment, four times each. Their results are marked on the thermometer scales in Figure 4:

- A precise set of results is grouped closely together.

- An accurate set of results will have a mean (average) close to the true value.

How do you get precise, repeatable data?

- You have to repeat your tests as often as necessary to improve repeatability.

- You have to repeat your tests in exactly the same way each time.

- You should use measuring instruments that have the appropriate scale divisions needed for a particular investigation. Smaller scale divisions have better resolution.

Making measurements
Using measuring instruments

There will always be some degree of uncertainty in any measurements made (WS3). You cannot expect perfect results. When you choose an instrument you need to know that it will give you the accuracy that you want (i.e., it will give you a true reading). You also need to know how to use an instrument properly.

Some instruments have smaller scale divisions than others. Instruments that measure the same thing, such as mass, can have different resolutions. The resolution of an instrument refers to the smallest change in a value that can be detected (e.g., a ruler with centimetre increments compared to a ruler with millimetre increments). Choosing an instrument with an inappropriate resolution can cause you to miss important data or make silly conclusions.

But selecting measuring instruments with high resolution might not be appropriate in some cases where the degree of uncertainty in a measurement is high, for example, judging when a wave in a ripple tank reaches the end of the tank (Topic 12.4). In this case an electronic timer measuring to within one thousandth of a second isn't going to improve the precision of the data collected.

WS3 Analysis and evaluation

Errors

Even when an instrument is used correctly, the results can still show differences. Results will differ because of a random error. This can be a result of poor measurements being made. It could also be due to not carrying out the method consistently in each test. Random errors are minimised by taking the mean of precise repeat readings, looking out for any outliers (measurements that differ significantly from the others within a set of repeats) to check again, or omit from calculations of the mean.

The **error** may be a **systematic error**. This means that the method or measurement was carried out consistently incorrectly so that an error was being repeated. An example could be a balance that is not set at zero correctly. Systematic errors will be consistently above, or below, the accurate value.

Presenting data
Tables
Tables are really good for recording your results quickly and clearly as you are carrying out an investigation. You should design your table before you start your investigation.

The range of the data
Pick out the maximum and the minimum values and you have the range. You should always quote these two numbers when asked for a range. For example, the range is between the lowest value in a data set, and the highest value. *Don't forget to include the units.*

Figure 6 *How you record your results will depend upon the type of measurements you are taking*

The mean of the data
Add up all of the measurements and divide by how many there are. You can ignore outliers in a set of repeat readings when calculating the mean, if found to be the result of poor measurement.

Bar charts
If you have a categoric independent variable and a continuous dependent variable then you should use a bar chart.

Line graphs
If you have a continuous independent and a continuous dependent variable then use a line graph.

Scatter graphs
These are used in much the same way as a line graph, but you might not expect to be able to draw such a clear line of best fit. For example, to find out if the melting point of an element is related to its density you might draw a scatter graph of your results.

Using data to draw conclusions
Identifying patterns and relationships
Now you have a bar chart or a line graph of your results you can begin looking for patterns. You must have an open mind at this point.

Firstly, there could still be some anomalous results. You might not have picked these out earlier. How do you spot an anomaly? It must be a significant distance away from the pattern, not just within normal variation.

A line of best fit will help to identify any anomalies at this stage. Ask yourself – 'do the anomalies represent something important or were they just a mistake?'

Secondly, remember a line of best fit can be a straight line or it can be a curve – you have to decide from your results.

The line of best fit will also lead you into thinking what the relationship is between your two variables. You need to consider whether the points you have plotted show a linear relationship. If so, you can draw a straight

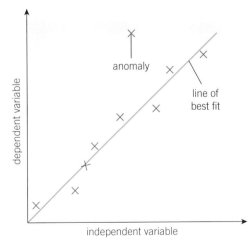

Figure 7 *A line of best fit can help to identify anomalies*

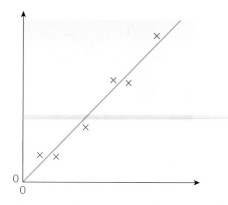

Figure 8 *When a straight line of best fit goes through the origin (0, 0) the relationship between the variables is directly proportional*

line of best fit on your graph (with as many points above the line as below it, producing a 'mean' line. Then consider if this line has a positive or negative gradient.

A directly proportional relationship is shown by a positive straight line that goes through the origin (0, 0).

Your results might also show a curved line of best fit. These can be predictable, complex or very complex. Carrying out more tests with a smaller interval near the area where a line changes its gradient will help reduce the error in drawing the line (in this case a curve) of best fit.

Drawing conclusions

Your graphs are designed to show the relationship between your two chosen variables. You need to consider what that relationship means for your conclusion. You must also take into account the repeatability and the reproducibility of the data you are considering.

You will continue to have an open mind about your conclusion.

You will have made a prediction. This could be supported by your results, it might not be supported, or it could be partly supported. It might suggest some other hypothesis to you.

You must be willing to think carefully about your results. Remember it is quite rare for a set of results to completely support a prediction or be completely repeatable.

Look for possible links between variables, remembering that a positive relationship does not always mean a causal link between the two variables.

Your conclusion must go no further than the evidence that you have. Any patterns you spot are only strictly valid in the range of values you tested. Further tests are needed to check whether the pattern continues beyond this range.

The purpose of the prediction was to test a hypothesis. The hypothesis can:

● be supported,

● be refuted, or

● lead to another hypothesis.

You have to decide which it is on the evidence available.

Making estimates of uncertainty

You can use the range of a set of repeat measurements about their mean to estimate the degree of uncertainty in the data collected.

For example, in a test to look at the descent of a parachute over a measured vertical distance, a student got the following results:

4.5 s, 4.9 s, 4.4 s, and 4.8 s.

The mean result = (4.5 + 4.9 + 4.4 + 4.8) ÷ 4 = 4.65 s

The range of these readings is 4.4 s to 4.9 s = 0.5 s.

So, a reasonable estimate of the uncertainty in the mean value would be half the range. In this case, the time taken would be ± 0.25 s.

You can include a final column in your table of results to record the 'estimated uncertainty' in your mean measurements.

The level of uncertainty can also be shown on a graph (Figure 9).

As well as this, there will be some uncertainty associated with readings from any measuring instrument. You can usually take this as:

- half the smallest scale division. For example, 0.5 mm on a metre ruler with millimetre division lines, or

- on a digital instrument, half the last figure shown on its display. For example, on a balance reading to 0.01 g the uncertainty would be ±0.005 g.

Anomalous results

Anomalies (or outliers) are results that are clearly out of line compared with others. They are not those that are due to the natural variation that you get from any measurement. Anomalous results should be looked at carefully. There might be a very interesting reason why they are so different.

If anomalies can be identified while you are doing an investigation, then it is best to repeat that part of the investigation. If you find that an anomaly is due to poor measurement, then it should be ignored.

Evaluation

If you are still uncertain about a conclusion, it might be down to the repeatability, reproducibility and uncertainty in your measurements. You could check reproducibility by: looking for other similar work on the Internet or from others in your class, getting somebody else, using different equipment, to redo your investigation (this occurs in peer review of data presented in articles in scientific journals), trying an alternative method to see if it results in you reaching the same conclusion.

When suggesting improvement that could be made in your investigation, always give your reasoning.

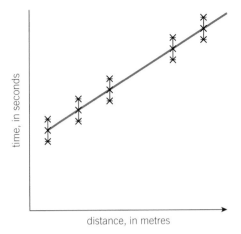

Figure 9 *Indicating levels of uncertainty. These are the results for the time it took for a parachute to fall a measured distance. For each distance, the test was repeated three times*

Study tip

The method chosen for an investigation can only be evaluated as being valid if it actually collects data that can answer your original question. The data should be repeatable and reproducible, and the control variables should have been kept constant (or taken into account if they couldn't be directly manipulated).

Synoptic link

See the Maths skills section to learn how to use SI units, prefixes and powers of ten for orders of magnitude, significant figures, and scientific quantities.

Glossary

Acceleration Change of velocity per second (in metres per second per second, m/s²).

Activity The number of unstable atoms that decay per second in a radioactive source.

Alpha radiation Alpha particles, each composed of two protons and two neutrons, emitted by unstable nuclei.

Alternating current Electric current in a circuit that repeatedly reverses its direction.

Amplitude The height of a wave crest or trough of a transverse wave from the rest position. For oscillating motion, the amplitude is the maximum distance moved by an oscillating object from its equilibrium position.

Atomic number The number of protons (which equals the number of electrons) in an atom. It is sometimes called the proton number.

Beta radiation Beta particles that are high-energy electrons created in, and emitted from, unstable nuclei.

Biofuel Any fuel taken from living or recently living materials, such as animal waste.

Boiling point Temperature at which a pure substance boils or condenses.

Braking distance The distance travelled by a vehicle during the time it takes for its brakes to act.

Carbon neutral A biofuel from a living organism that takes in as much carbon dioxide from the atmosphere as is released when the fuel is burned.

Carrier waves Waves used to carry any type of signal.

Charge-coupled device (CCD) An electronic device that creates an electronic signal from an optical image formed on the CCD's array of pixels

Closed system An object or a group of objects for which the total energy is constant.

Compression Squeezing together.

Conservation of energy Energy cannot be created or destroyed.

Conservation of momentum In a closed system, the total momentum before an event is equal to the total momentum after the event. Momentum is conserved in any collision or explosion, provided no external forces act on the objects that collide or explode.

Contrast medium An X-ray absorbing substance used to fill a body organ so the organ can be seen on a radiograph.

Count rate The number of counts per second detected by a Geiger counter.

Deceleration Change of velocity per second when an object slows down.

Density Mass per unit volume of a substance.

Diode A non-ohmic conductor that has a much higher resistance in one direction (its reverse direction) than in the other direction (its forward direction).

Direct current Electric current in a circuit that is in one direction only.

Directly proportional A graph will show this if the line of best fit is a straight line through the origin.

Displacement Distance in a given direction.

Dissipation of energy The energy that is not usefully transferred and stored in less useful ways.

Driving force Force of a vehicle that makes it move (sometimes referred to as motive force).

Earth wire The wire in a mains cable used to connect the metal case of an appliance to earth.

Echo Reflection of sound that can be heard.

Efficiency Useful energy transferred by a device ÷ total energy supplied to the device.

Elastic A material is elastic if it is able to regain its shape after it has been squashed or stretched.

Electromagnet An insulated wire wrapped round an iron bar that becomes magnetic when there is a current in the wire.

Electromagnetic spectrum The continuous spectrum of electromagnetic waves.

Electromagnetic waves Electric and magnetic disturbances that transfer energy from one place to another.

Electrons Tiny negatively charged particles that move around the nucleus of an atom.

Errors Sometimes called uncertainties.

Extension The increase in length of a spring (or a strip of material) from its original length.

Fleming's left-hand rule A rule that gives the direction of the force on a current-carrying wire in a magnetic field according to the directions of the current and the field.

Force A force (in newtons, N) can change the motion of an object.

Free body force diagram A diagram that shows the forces acting on an object without any other objects or forces shown.

Freezing point The temperature at which a pure substance freezes.

Frequency The number of wave crests passing a fixed point every second.

Frequency of an alternating current The number of complete cycles an alternating current passes through each second. The unit of frequency is the hertz (Hz).

Frequency of oscillating motion Number of complete cycles of oscillation per second, equal to 1 ÷

the time period. The unit of frequency is the hertz (Hz).

Friction The force opposing the relative motion of two solid surfaces in contact.

Fuse A fuse contains a thin wire that melts and cuts the current off if too much current passes through it.

Gamma radiation Electromagnetic radiation emitted from unstable nuclei in radioactive substances.

Geothermal Energy that comes from energy released by radioactive substances deep within the Earth.

Gradient (of a straight line graph) Change of the quantity plotted on the y-axis divided by the change of the quantity plotted on the x-axis.

Gravitational field strength, g The force of gravity on an object of mass 1 kg (in newtons per kilogram, N/kg). It is also the acceleration of free fall.

Half-life Average time taken for the number of nuclei of the isotope (or mass of the isotope) in a sample to halve.

Hooke's law The extension of a spring is directly proportional to the force applied, as long as its limit of proportionality is not exceeded.

Induced magnetism Magnetisation of an unmagnetised magnetic material by placing it in a magnetic field.

Inertia The tendency of an object to stay at rest or to continue in uniform motion.

Infrared radiation Electromagnetic waves between visible light and microwaves in the electromagnetic spectrum.

Input energy Energy supplied to a device.

Internal energy The energy of the particles of a substance due to their individual motion and positions

Inverse proportionality This is where two variables are related such that making one variable n times bigger causes the other one to become n times smaller (e.g. doubling one quantity causes the other one to halve).

Ion A charged atom or molecule.

Ionisation Any process in which atoms become charged.

Irradiation Exposure of an object to ionising radiation.

Isotopes Atoms with the same number of protons and different numbers of neutrons.

Kilo-watthour (kWh) The energy in electricity supplied to a 1 kW electrical device in 1 hour.

Latent heat The energy transferred to or from a substance when it changes its state

Light-depending resistor (LDR) A resistor whose resistance depends on the intensity of the light incident on it.

Light-emitting diode (LED) A diode that emits light when it conducts.

Limit of proportionality The limit for Hooke's law applied to the extension of a stretched spring.

Line of action The line along which a force acts.

Line of force Line in a magnetic field along which a magnetic compass points – also called a magnetic field line.

Live wire The mains wire that has a voltage that alternates in voltage (between +325 V and 325 V in Europe).

Load The weight of an object raised by a device used to lift the object, or the force applied by a device when it is used to shift an object.

Longitudinal waves Waves in which the vibrations are parallel to the direction of energy transfer.

Magnetic field The space around a magnet or a current-carrying wire.

Magnetic field line Line in a magnetic field along which a magnetic compass points – also called a line of force.

Magnetic flux density A measure of the strength of the magnetic field defined in terms of the force on a current-carrying conductor at right angles to the field lines.

Magnitude The size or amount of a physical quantity.

Mass The quantity of matter in an object – a measure of the difficulty of changing the motion of an object (in kilograms, kg).

Mass number The number of proton and neutrons in a nucleus.

Mechanical wave Vibration that travels through a substance.

Melting point Temperature at which a pure substance melts or freezes (solidifies).

Microwaves Electromagnetic waves between infrared radiation and radio waves in the electromagnetic spectrum.

Momentum This equals mass (in kg) × velocity (in m/s).

Motor effect When a current is passed along a wire in a magnetic field, and the wire is not parallel to the lines of the magnetic field, a force is exerted on the wire by the magnetic field.

National Grid The network of cables and transformers used to transfer electricity from power stations to consumers (i.e., homes, shops, offices, factories, etc.).

Neutral wire The wire of a mains circuit that is earthed at the local substation so its potential is close to zero.

Neutrons Uncharged particles of the same mass as protons. The nucleus of an atom consists of protons and neutrons.

Newton's first law of motion If the resultant force on an object is zero, the object stays at rest if it is stationary, or it keeps moving with the same speed in the same direction.

Newton's second law of motion The acceleration of an object is proportional to the resultant force on the object, and inversely proportional to the mass of the object.

Newton's third law of motion When two objects interact with each other, they exert equal and opposite forces on each other.

Nuclear fuel Substance used in nuclear reactors that releases energy due to nuclear fission.

Nucleus Tiny positively charged object composed of protons and neutrons at the centre of every atom.

Ohm's law The current through a resistor at constant temperature is directly proportional to the potential difference across the resistor.

Optical fibre Thin glass fibre used to transmit light signals.

Oscillate Move to and fro about a certain position along a line.

Oscilloscope A device used to display the shape of an electrical wave.

Parallel Components connected in a circuit so that the potential difference is the same across each one.

Parallelogram of forces A geometrical method used to find the resultant of two forces that do not act along the same line.

Physical change A change in which no new substances are produced.

Plugs A plug has an insulated case and is used to connect the cable from an appliance to a socket.

Potential difference A measure of the work done or energy transferred to the lamp by each coulomb of charge that passes through it. The unit of potential difference is the volt (V).

Power The energy transformed or transferred per second. The unit of power is the watt (W).

Protons Positively charged particles with an equal and opposite charge to that of an electron.

Radiation dose Amount of ionising radiation a person receives.

Radio waves Electromagnetic waves of wavelengths greater than 0.10 m.

Radioactive contamination The unwanted presence of materials containing radioactive atoms on other materials.

Rarefaction Stretched apart.

Reactor core The thick steel vessel used to contain fuel rods, control rods and the moderator in a nuclear fission reactor.

Reflection The change of direction of a light ray or wave at a boundary when the ray or wave stays in the incident medium.

Refraction The change of direction of a light ray when it passes across a boundary between two transparent substances (including air).

Renewable energy Energy from natural sources that is always being replenished so it never runs out

Resistance Resistance (in ohms, Ω) = potential difference (in volts, V) ÷ current (in amperes, A).

Resultant force A single force that has the same effect as all the forces acting on the object.

Risk The likelihood that a hazard will actually cause harm.

Scalar A physical quantity, such as mass or energy, that has magnitude only (unlike a vector which has magnitude and direction).

Series Components connected in a circuit in such a way that the same current passes through them.

Solar heating panel Sealed panel designed to use sunlight to heat water running through it.

Solenoid A long coil of wire that produces a magnetic field in and

around the coil when there is a current in the coil

Specific heat capacity Energy needed to raise the temperature of 1 kg of a substance by 1 °C.

Specific latent heat of fusion Energy needed to melt 1 kg of a substance with no change of temperature.

Specific latent heat of vaporisation Energy needed to boil away 1 kg of a substance with no change of temperature.

Speed The speed of an object (metres per second) = distance moved by the object (metres) ÷ time taken to move the distance travelled (seconds).

Split-ring commutator Metal contacts on the coil of a direct current motor that connects the rotating coil continuously to its electrical power supply.

Spring constant Force per unit extension of a spring.

Static electricity Electric charge stored on insulated objects.

Step-down transformer Electrical device that is used to step-down the size of an alternating potential difference.

Step-up transformer Electrical device used to step-up the size of an alternating potential difference.

Stopping distance The distance travelled by the vehicle in the time it takes for the driver to think and brake.

Systematic errors Cause readings to be spread a value other than the true value, due to results differing from the true value by a consistent amount each time a measurement is made.

Tangent A straight line drawn to touch a point on a curve so it has the same gradient as the curve at that point.

Terminal velocity The velocity reached by an object when the drag force on it is equal and opposite to the force making it move.

Thermal conductivity Property of a material that determines the energy transfer through it by conduction.

Thermistor A resistor whose resistance depends on the temperature of the thermistor.

Thinking distance The distance travelled by the vehicle in the time it takes the driver to react.

Three-pin plug A three-pin plug has a live pin, a neutral pin, and an earth pin.

Transformer Electrical device used to change an (alternating) voltage. See also Step-up transformer and Step-down transformer.

Transmission A wave passing through a substance.

Transparent object Object that transmits all the incident light that enters the object.

Transverse wave A wave where the vibration is perpendicular to the direction of energy transfer.

Ultraviolet radiation Electromagnetic waves between visible light and X-rays in the electromagnetic spectrum.

Useful energy Energy transferred to where it is wanted in the way that is wanted.

Vector A vector is a physical quantity, such as displacement or velocity, that has a magnitude and a direction (unlike a scalar which has magnitude only).

Velocity Speed in a given direction (in metres/second, m/s).

Vibrate Oscillate (move to and fro) rapidly about a certain position.

Wasted energy Energy that is not usefully transferred.

Wave speed The distance travelled per second by a wave crest or trough.

Wavelength The distance from one wave crest to the next.

Weight The force of gravity on an object (in newtons, N).

White light Light that includes all the colours of the spectrum.

Work The energy transferred by a force. Work done (joules, J) = force (newtons, N) × distance moved in the direction of the force (metres, m).

X-rays Electromagnetic waves smaller in wavelength than ultraviolet radiation produced by X-ray tubes.

Index

Appendix 1: Physics equations

You should be able to remember and apply the following equations, using SI units, for your assessments.

Word equation	Symbol equation
weight = mass × gravitational field strength	$W = mg$
force applied to a spring = spring constant × extension	$F = ke$
acceleration = $\dfrac{\text{change in velocity}}{\text{time taken}}$	$a = \dfrac{\Delta v}{t}$
Ⓗ momentum = mass × velocity	$p = mv$
gravitational potential energy = mass × gravitational field strength × height	$E_p = mgh$
power = $\dfrac{\text{work done}}{\text{time}}$	$P = \dfrac{W}{t}$
efficiency = $\dfrac{\text{useful power output}}{\text{total power input}}$	
charge flow = current × time	$Q = It$
power = potential difference × current	$P = VI$
energy transferred = power × time	$E = Pt$
density = $\dfrac{\text{mass}}{\text{volume}}$	$\rho = \dfrac{m}{V}$
work done = force × distance (along the line of action of the force)	$W = Fs$
distance travelled = speed × time	$s = vt$
resultant force = mass × acceleration	$F = ma$
kinetic energy = 0.5 × mass × (speed)^2	$E_k = \dfrac{1}{2}mv^2$
power = $\dfrac{\text{energy transferred}}{\text{time}}$	$P = \dfrac{E}{t}$
efficiency = $\dfrac{\text{useful output energy transfer}}{\text{total input energy transfer}}$	
wave speed = frequency × wavelength	$v = f\lambda$
potential difference = current × resistance	$V = IR$
power = current^2 × resistance	$P = I^2R$
energy transferred = charge flow × potential difference	$E = QV$

You should be able to select and apply the following equations
from the Physics equation sheet.

Word equation	Symbol equation
(final velocity)^2 − (initial velocity)^2 = 2 × acceleration × distance	$v^2 - u^2 = 2as$
elastic potential energy = 0.5 × spring constant × extension^2	$E_e = \dfrac{1}{2}ke^2$
period = $\dfrac{1}{\text{frequency}}$	
⊕ force on a conductor (at right angles to a magnetic field) carrying a current = magnetic flux density × current × length	$F = BIl$
change in thermal energy = mass × specific heat capacity × temperature change	$\Delta E = mc\Delta\theta$
thermal energy for a change of state = mass × specific latent heat	$E = mL$